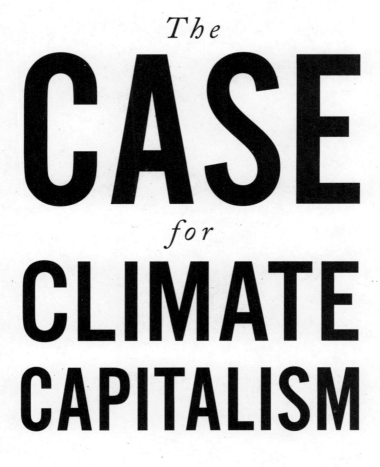

The

CASE

for

CLIMATE

CAPITALISM

CASE

for

CLIMATE

CAPITALISM

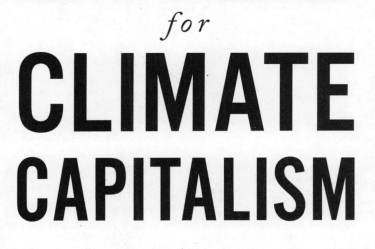

*Economic Solutions for a
Planet in Crisis*

TOM RAND

Copyright © Tom Rand, 2020

Published by ECW Press
665 Gerrard Street East
Toronto, Ontario, Canada M4M 1Y2
416-694-3348 / info@ecwpress.com

Cover design: Michel Vrana

LIBRARY AND ARCHIVES CANADA CATALOGUING
IN PUBLICATION

Title: The case for climate capitalism : economic solutions for a planet in crisis / Tom Rand.

Names: Rand, Tom, 1967– author.

Description: Includes bibliographical references and index.

Identifiers: Canadiana (print) 20190177950 | Canadiana (ebook) 20190177985

ISBN 9781770415232 (hardcover)
ISBN 9781773055107 (PDF)
ISBN 9781773055091 (EPUB)

Subjects: LCSH: Environmental economics. | LCSH: Capitalism—Environmental aspects. | LCSH: Climatic changes—Economic aspects. | LCSH: Social responsibility of business.

Classification: LCC HC79.E5 R36 2020 | DDC 338.9/27—dc23

The publication of *The Case for Climate Capitalism* has been funded in part by the Government of Canada. *Ce livre est financé en partie par le gouvernement du Canada.* We acknowledge the contribution of the Government of Ontario through the Ontario Book Publishing Tax Credit, and through Ontario Creates for the marketing of this book.

PRINTED AND BOUND IN CANADA

PRINTING: FRIESENS   5   4   3   2   1

For Rupert, of course, with love unbounded.

And for the brave Greta Thunberg:

"Adults keep saying, 'We owe it to the young people to give them hope.' But I don't want your hope. I don't want you to be hopeful. I want you to panic.

"I want you to feel the fear I feel every day. And then I want you to act.

"I want you to act as you would in a crisis. I want you to act as if our house is on fire. Because it is."

— SPEAKING AT THE WORLD ECONOMIC FORUM
IN DAVOS, SWITZERLAND, 2019

CONTENTS

"Man achieves civilization not as a result of superior biological endowment or geographical environment, but as a response to a challenge in a situation of special difficulty which rouses him to make a hitherto unprecedented effort."
— ARNOLD TOYNBEE[1]

PREFACE

Partway through writing this book, I discovered I was going to become a father. His mother and I were both surprised as it proved wrong a cadre of medical experts. The usual panic ensued and faded. It was obvious my life would change in lots of ways, many unexpected, mostly good, some inconvenient. But I knew I would handle it; after all, the same thing's happened to billions of others for thousands of years! It's part of being human. But one deep-seated fear simply won't go away. It may not be something I can ever handle with grace. It's partly why this kid thing hadn't already happened.

Climate change terrifies me. Jokes about apocalyptic landscapes are as much confession as humor to climate nerds. It's not just me — anyone who's been in the carbon kitchen comes out in an uncomfortable emotional state.[i] Each person deals with those demons in his or her own way. Eric Holthaus — a meteorologist and former columnist for the *Wall*

---

i   There's a new term, "climate trauma," used to describe the increased psychological tension resulting from being immersed in scary climate data. See "As Droughts, Floods, Die-Offs Proliferate, 'Climate Trauma' a Growing Phenomenon" at newsecuritybeat.org, Sept. 9, 2015.

Street Journal who writes about the impacts of global climate change —
came out publicly about the way his work affected his mental health.
He's not the only one. By drawing attention to his own struggles, he
normalized a conversation many of us want to have.

What's that got to do with the kid? The prospect of hitting catastrophic tipping points in the next few decades brings more than sleepless
nights — it makes having a kid a complex moral question.[ii] First, there's
the issue of bringing yet another person to our profligate emissions party.
It's uncomfortable (to put it mildly) to say climate change is as much
a population issue[iii] as an environmental one. Having kids is no longer
morally benign — or even good — as it was in previous generations. But
for me, it's not just the population thing. After all, someone's got to have
them, and heck, I'm only having one!

My deep unease is because I know — absent a near-miraculous
change in political will — my kid is likely to grow up in an age driven
by scarcity and conflict.[iv] I can't forget what I've learned about climate
risk. There are good reasons most sane people tune out increasingly shrill
warnings from the scientific and security communities. They are deeply
unsettling, even anxiety-inducing.

One way I've been keeping my own climate demons at bay was to
believe I was one step removed from that possible future. I'm connected
by love to an extended network of friends and family, of course, but with
no kids of my own, I could fool myself into seeing climate risk as a philosophical problem or interesting anthropological challenge rather than
a direct emotional threat. I'd live to see some effects, but by the time
things got catastrophic — say 2040 or 2050 — I'd be long gone with

---

ii   Groups like Conceivable Future are emerging to provide support to climate activ-
      ists dealing with the very real moral question of having kids.

iii   A 2010 study found slowing population growth could take care of 20 to 25 percent
      of the carbon emissions cuts required to avoid catastrophic tipping points by mid-
      century. O'Neill, Brian C., et al. "Global Demographic Trends and Future Carbon
      Emissions," *Proceedings of the National Academy of Sciences* 107: 17521–17526, 2010.

iv   It's true many people today live in that sort of misery. But with empathy and good
      governance comes progress.

no real stakes in play. No more. My kid will be an adult then. Without some sharp turns in our economic system, he'll almost certainly see some degree of breakdown in social order.[v] So now I'm in way deeper than I'd planned, and it's uncomfortable. Those demons have come back with a vengeance.

Days after discovering I was to be a dad, I attended the 2016 Planetary Security Conference in The Hague. Military brass and security experts, one after another, confirmed what I already knew: the degree of social upheaval coming our way is mind-numbing. The language they used was somber: substantial risks of societal collapse; catastrophic effects; destabilization. That very year, every refugee hitting European shores came from water-stressed countries. The Syrian war was exacerbated by the longest and most severe drought in memory. The razor-wire fence around Bangladesh, backed by shoot-to-kill soldiers, is a dark premonition of what will happen when that crowded country begins to flood and the population looks to escape. When Russia temporarily shut down grain exports during a severe drought in 2012, the cost of food around the world skyrocketed. The resulting riots kicked off the Arab Spring.

The military community doesn't see climate change as a simple set of individuated physical events, like droughts or storms, but as a "threat multiplier" that amplifies social tensions in ways we are just beginning to appreciate. Water stress in one place brings risk elsewhere: those Syrian refugees triggered a resurgence of the far right throughout Europe. And food markets are global; leaders cannot see their populations starve without resorting to what military might they have to get the food and water they need.[vi] Refugees by the tens or even hundreds of millions will not be contained in camps or by walls. Political tensions build in complex, interactive ways. We do not live in isolation. Our individual security is predicated on collective security. My kid can't hide, even in a place like Canada.

We're seeing now in the U.S. what can happen when people get nervous. Populist demagogues who play to those fears find room to

---

v An estimated 700 million climate refugees will be on the move by then.

vi See Gwynne Dyer's *Climate Wars*.

breathe. Instead of banding together, walls — both real and metaphoric — go up. If people are nervous now because of recent mild shocks to accepted norms of economic progress, they'll be terrified later. And, of course, a climate denier currently occupies the highest office on the planet. Things can go the other way. People in times of stress often come together and elect leaders of compassion and vision. But Trump's win and early actions as president show that's hardly a given. Most alarming is democracy's massive overreaction to relatively tiny levels of personal insecurity compared to what's coming. That reasoned debate and the rule of law in the United States are under threat today bodes ill for my kid and his peers.

My kid arrived just as the storms gather intensity. It was a hard week in The Hague. And a difficult year. It was tempting — and perfectly rational — to look the other way, to hibernate in a home filled with love, to leave the worry to others. But I can't look the other way. I can't give up. If for no other reason, I want that kid to know I tried.

We can't solve climate change, but we can still head off the worst. That's my life's work now more than ever. I've no choice but to dig deeper for solutions. This book is one of my shovels. So is my clean energy technology fund, ArcTern Ventures.[vii] So is my work at the Toronto-based MaRS Discovery District.[viii] So, too, are my weekly debates on CBC News Network and elsewhere with right-wing libertarians who can't yet acknowledge that we should even try to combat the coming storm. Whatever happens, that kid will know I did my best. I won't choose the luxury of looking the other way, of getting lost in domestic life.

Climate change will, more than anything else, define the economic and social conditions of my kid's generation. That's true for Canadians, Pakistanis, and Africans. It holds for rich and poor, men and women, artists and venture capitalists. This is true whether or not we limit warming

---

vii   We invest private capital, including my own, into companies building "cleantech," technology that can beat fossil fuels at their own game.

viii   MaRS Cleantech advises start-ups, connects them to capital and top-tier talent, and accelerates them into global markets. MaRS and ArcTern work hand in hand.

to relatively safe levels. To mitigate climate risk means decades of radical change to the economy, change we have not yet begun in earnest. Absent that effort, we face centuries of economic uncertainty and social unrest. We either make the kind of change that rights the ship as best we can, or we live with disruptive, dislocating, and damaging changes to the economy and planet. Either way, climate now dominates economic and social life. Some form of Climate Capitalism is inevitable.

I was once an idealist and believed the key to living a moral life or solving a complex problem was to find the right core ideas — the right ideology — and commit to them. Solutions required only their dedicated application. Often, they could be narrowed down to a few simple principles. It's an attractive idea that has seduced thinkers for ages. A few years of advanced philosophy convinced me otherwise. Entrenched positions on the left and right in the climate debate are extensions of simple core beliefs, and they bring more discord than solutions to such a complex problem.

Pragmatism is a technical term in philosophy, but it means largely what you think it might. Ideologies are true only insomuch as they are useful, not because they are internally consistent or logical. The meaning of a statement is bound up in its consequences, not in some abstracted set of definitions. Unpractical ideas, while perhaps ideologically satisfying or intellectually beautiful, are to be rejected. The best we can do is to approach the world from whatever position allows us to understand it at that time and in that place, gives us an ability to solve problems we find relevant, and moves things forward in a way that makes intuitive sense. We take what works, throw out what doesn't, and muddle on.

Our shared human world is a messy place that can't be captured by a single, coherent view. If an ideology (take your pick: physical, moral, social, political, or economic) is coherent, it misses much of what we take to be obvious. Capturing what's obvious comes at the cost of coherence and simplicity. British philosopher Bertrand Russell tried to ground mathematics in pure logic, but Kurt Gödel burned that effort to the ground. Moral theorists have never really managed to come up with anything much better than the Golden Rule. And Thomas Kuhn showed even science can't operate independent of psychological and social systems. Common sense — our capacity to make sense of the world in deeply intuitive ways — always grounds our more theoretical ways of understanding. Pragmatism was a reaction against philosophical ideologies that have to do with meaning, truth, and language. But its lessons apply here.

Climate change defies the traditional divisions between left and right. The response from the far left, led by Naomi Klein, targets market forces, economic growth, and capitalism itself as the enemy. Yet climate solutions need all three. The far right, dominated by market fundamentalists like FOX News and the Koch brothers, view unfettered markets, unlimited growth, and unregulated capitalism as unassailable foundations of the twenty-first century. But that view is incompatible with a livable planet. The simple ideologies of left and right are unhelpful in trying to solve this problem. It's time to let them go. Common sense provides a better basis for climate solutions than political or ideological preference.

Climate Capitalism is a pragmatic response to a messy problem. To rewire our economy in time to head off disaster, the left and right need to throw out a bunch of comfortable assumptions. The idea that we're going to jettison capitalism itself is as absurd as it sounds. We need high finance. And market forces. Yet both must be tamed. Unbridled market forces make for great toys and factories but will take us straight off the climate cliff. Not everything can be valued in money or commerce. Some things have worth that can't be contained in a spreadsheet: the human spirit, our place in the world, values and ethics, our planet itself.

The current intellectual trend claims all of human activity can be captured in value-free quantitative analysis. That view is false. We can't speak to the climate issue without the deep, reflective language of moral philosophy. Intergenerational justice is not measured by what economists call the "discount rate." The value of nature — the Amazonian rainforest, biodiversity, healthy watersheds — is not captured by estimating their monetary value. Our deeper qualitative concerns must bend (not bend *to*) the forces of commerce to be effective. Commerce is the tool, human values the force — not vice versa.

Yet this book is all about economics: money, trade agreements, discount rates, capital markets, entrepreneurs. Undoubtedly, my environmental friends won't like the whole-hearted embrace of that language. And many business colleagues will bristle at the call for what they see as radical interference in those markets and a value-first approach to weighing costs and benefits. That may be a sign I've got something right. In any good negotiation, both sides will feel like they lost. But both also win.

Things will get nasty in the climate debate as our world continues to get hotter. There will be fights — not just over ideas but water, food, land, and money. But one thing we can't fight about anymore is which economic system occupies the high ground. The left and right, the business community and environmentalists, bankers and activists must together reclaim capitalism and force profits to align with the planet. We must retool our laws and institutions to reflect our collective long-term security. We can worry about who occupies the moral high ground later.

Climate is especially tough because we're all complicit. It's not so much about emissions as about people and what we do. It's deeply personal, not just political or theoretical. I was fortunate enough to be supported in writing this book by a Rockefeller Foundation Residency, which put me in a gorgeous villa in Bellagio, Italy, for a month. I didn't walk there, I flew. In one flight, I blew out more carbon than the average Ethiopian does in a year. My kid is probably the largest single carbon emission I'll ever be responsible for. We all live with the hypocrisy of being part of a problem we're trying to solve.

Hypocrisy reveals both value and human fallibility. Having values to contravene is what makes hypocrisy possible. It's easy to criticize climate advocates who use fossil fuels. I remember as a kid watching on TV a Greenpeace boat with a large "Go Solar" flag harass an oil tanker. I asked my dad why the Greenpeace boat used oil. "Do you think they could catch up to the tanker using sails?" he asked. We don't need to live in a cave to ask for change. But we do need to take what action we can. Each of us draws the line where it feels right.

My kid will undoubtedly ask me tough questions like, "What did you do when things started getting hot?" This book is one part of that answer.

We can't solve climate change in the sense that we can make it go away. It's far too late for that. But we can tame this wicked problem if we leave our dogma at the door.

March 2019

P.S. That kid is now here. I call him Rupe, short for Rupert.

INTRODUCTION

You don't have to be a card-carrying socialist, or even an environmentalist, to want radical action on climate. That's just good sense. Nor must you be a greed-is-good Wall Street titan to believe market forces, profit, and capital are critical pieces of that action. That, too, is good sense. Rewiring existing economic systems to lower climate risk could not be a more centrist notion. Yet it's increasingly seen as radical: by the new left because it fails to reject capitalism outright, and by the new right because it implies massive market interventions. Our respective dogmas prevent the consensus needed to build a new economy that will get us safely through the century. These entrenched positions preclude finding workable solutions. But there is no contradiction between capitalism and forceful public action. This book is a plea to reengage on that centrist idea: Climate Capitalism as a radical rethinking of ideas as old as industrial civilization itself.

There are many conflicting positions on climate disruption. Getting bogged down in them isn't getting us to a collective solution. That much is clear. But by exploring the key — and often opposing — positions, we can take what's true and sensible from each to show us a new way forward. Climate Capitalism is a set of pragmatic solutions to mitigate the single greatest risk we face this century. It's also a plea to put the bickering behind us and put those solutions in place quickly and aggressively. There's no single perfect answer or magic bullet — but there is a clear path to the radical action we need. When your house is on fire, it does no good to argue about who started it, how hot it might get, or whether the insurance will pay; what you need is a good, fat fire hose. Climate Capitalism is that hose.

We need to move on from lobbing salvos from our preferred ideological camp. Adhering to a kind of moral purity on this issue is a liability, not an asset, and undermining others' position is wasted energy, time, and social capital. If we don't come together on this urgently to find solutions, everything we know will go up in smoke. Each view has merit, and each has flaws. Climate Capitalism urges us to find common ground to tackle this most pressing and confounding problem for good — our good.

The reason the left sees capitalism itself as the bogeyman and blames it for where we are with the environment is because of businesspeople out there like the Koch brothers — powerful industrialists, wealthy beyond imagination, working their hearts out in a way that will destroy the world. I have no idea whether they're sociopaths or not, but they might as well be. To protect old ways of doing business and make their pile of gold even bigger, these Scrooge McDucks go to the wall to prevent action on climate. In the guise of an extreme neoconservative view of markets (which has been discredited by most experts), they hijack politics by way of a corrupt political system and distort modern capitalism. Others, like Rupert Murdoch, have built media empires that poison debate by dumbing down the conversation and coloring public perceptions of climate risk. The power accrued to these dangerous oligarchs is not a trivial flaw in capitalism. Instead, it speaks to the need for strong public institutions and oversight to prevent this kind of dark money from perverting democracy itself.

In fact, the new economy is being built today by some inspiring figures, including entrepreneurs who work within the capitalist system to help solve our climate challenges. John Paul Morgan is one of the smartest and most capable people you're likely to meet. His intelligence is buzzy, like a vibrating field you sense around his head, as he dives into almost any topic — physics, economics, global markets, technology, finance. He's also likely the most well-meaning and moral person you'll come across in the business community. He built hospitals in Africa for Doctors Without Borders. He adopted his three-legged dog after it was hit by a car in Colombia simply because nobody else would. That's John Paul. And he's why I invested in his company.

He founded Morgan Solar alongside his intense and gregarious father, Eric, and older brother and longtime mentor, Nicolas. John Paul left Africa because Eric challenged him to make an even bigger difference in the world. He figured he could meet that challenge by making solar power cheaper than coal. Morgan Solar is about to do just that. There are lots of entrepreneurs like John Paul working their hearts out to raise money, prove technology, build factories, and — yes — save the world.

John Paul is the anti-Koch, and people like him are the reason I still believe capitalism might, just might, be able to respond to the climate crisis. John Paul may be a moral light, but he's not naïve. Like his better-known American counterpart Elon Musk, he knows the best way to scale up energy to empower the poor, reduce greenhouse gas (GHG) emissions, and energize the world is to build a world-class, massively successful clean energy company. That means wooing investors, building global partnerships, competing head-to-head with coal in the open market, and making money doing it. We can't build that vision, one shared by environmentalists on the left, absent a market economy.

Capitalism isn't John Paul, nor is it the Koch brothers. It's neither good nor bad. It is shaped by local culture and politics. Far from being a single economic system, it comes in many flavors: Wall Street's finance capitalism; Russia's rough-and-tumble anarcho-capitalism; Indonesia's crony capitalism; China's state capitalism; the protectionism that dominated trade in the middle part of the twentieth century and the globalized markets of the twenty-first century; activist public sectors like Roosevelt's New Deal; the unfettered markets of neoconservatives.

Capitalism is how we've done business since feudal times — an ongoing attempt to use markets as an organizing principle of our daily transactions. As economies have become more complex, the role of distributed decision-making in the form of market forces has become more crucial. The few ill-fated exceptions[i] prove the rule that market forces will always play a critical organizing role in complex economic activity. The battle between left and right is mainly about the degree of freedom under which those market forces operate. Climate Capitalism is no more than a new set of rules, borrowing from both the left and the right, to make our economic system responsive to the climate threat.

The tide is turning. The Kochs of the world are on the wane, and entrepreneurs like John Paul are on the rise. Technology is unstoppable once it gets going. Its sweeping changes are clear in hindsight: the engines of the industrial revolution; the microchips of the data revolution; the genetic and stem cell manipulations that herald a new era of human health. Similarly, clean energy technology, or cleantech, is igniting an energy revolution. Instead of digging up and burning stuff buried in the ground, technology will harness and store endless supplies of renewable energy. Cars will not use internal combustion engines for much longer. The economic dynamism at the dawn of the twenty-first century feels unstoppable. The potential of the creative class seems unbound. The human spirit runs hot.

But so does the planet. Lurking behind today's technological optimism is the biggest, baddest dragon of modern times. Climate disruption threatens our civilization in ways the public is only just beginning to understand. Already we're seeing waters rise, storms grow, rainfall patterns change, and droughts intensify as the atmosphere heats up. The greenhouse gases we've already emitted trap the energy equivalent of hundreds of thousands of atomic bombs in the atmosphere every single day. It's not so much about heat, but energy. We're destabilizing the atmosphere, shifting it into a new state.

In this book, from this point forward, I will use the term "climate disruption" instead of the more broadly used "climate change," to be more accurate about what we are really facing. In an early interview with

---

i   Venezuela under Nicolás Maduro is melting into economic ruin. North Korea's miserably poor population has never seen anything but economic ruin. The Soviet Union's collapse, while evidence of the long-term instability of command-and-control economies, predicated a triumphalist overreaction by neoconservatives in the west.

Lesley Stahl, U.S. President Donald Trump implied that while the climate may be changing, it may change back. I won't spend much time here relitigating the sheer scientific ignorance expressed by the most powerful politician on the planet, but by using the term climate disruption[ii] I wish to emphasize *there is no going back*. The changes we are making are permanent disruptions to the ecosystem in which our civilization evolved, and to which our many layers of infrastructure — from food to buildings — are carefully attuned. We are heading down a dangerous one-way alley.

These are the early stages of a very risky experiment, one that may already be spinning out of our control. The precautionary principle says we have a duty to prevent harm even when all the evidence isn't in. Climate risk is more perilous than the subprime financial turmoil, China debt crises, terrorism, or the Middle East's ongoing oil and religious wars. Global food security, the gift of the last Green Revolution, is no longer a given. A more turbulent atmosphere is the ultimate creative destruction, wrought by an economy seemingly unable to curb its excesses. Unchecked emissions guarantee a catastrophic outcome, even if the timing and details are uncertain. We're way past the precautionary principle now.

Climate risk is qualitatively different from other threats. Once it spins out of control, there is no coming back, no second chance. Economic crises, security breaches, and terrorist activity are frightening and destabilizing. But they're temporary. Humanity always bounces back, often quickly. A climate crisis is effectively permanent. The ongoing instability of an overheated planet will last longer than we've had an industrial civilization. As a threat multiplier, it exacerbates *all* other crises: food, security, economic, social. The only risks comparable in outcome are those of a nuclear winter or an aberrant asteroid. This really is our endgame. We have to get it right. Failure is forever.

Two competing forces — the degree of freedom provided to markets and increasing climate risk — have forced open a cultural schism, bound to get wider as the planet warms. On one side stand those distrustful of

---

ii   To my knowledge this was first proposed by the Obama administration's science advisor John Holdren.

markets and economic growth. Primary spokespersons Naomi Klein and the Pope argue capitalism is the root cause of climate disruption. Only by ridding ourselves of its brutal specter might we save ourselves from the coming storms. On the other side stand modern neoconservatives, chanting something akin to the "What, me worry?" mantra of *MAD* magazine's Alfred E. Neuman. Scornful of those who criticize the market's efficiency, they wind themselves in knots defending business as usual by denying climate risk. Each side is guilty of taking a far too narrow view of capitalism.

Somewhere in the untenable middle, unable to straddle the widening gap much longer, stands what has been until now a fairly quiet group. Centrists — including many business leaders, economists, and other purveyors of opinion on our editorial pages — finally understand we're up the climate creek and have begun to assert themselves. Yet this group remains reluctant to endorse the kind of severe change in direction required to temper the coming storms. Climate risk may be real, the thinking goes, but to tamper radically with the heart of our energy and economic systems is riskier still.

This centrist position cannot hold. The cruel math of emission levels and climate risk is simply too relentless. If you're a moderate on action, you get extreme climate disruption. If you want moderate climate risk, you must endorse extreme action. There is simply no middle ground left.

The fruits of economic development are impossible to ignore. Those who were born in the latter half of the twentieth century in Canada practically won the lottery — it was one of the greatest times and places to come into being in human history. The massive economic expansion of the late twentieth century delivered employment and wealth. New technology brought better health care, cleaner air, and an endless array of new toys. Globalization brought down their cost. Food security seemed guaranteed. Cracks occasionally appeared as energy demand soared and the geopolitics of oil got nasty, but resource scarcity was something that happened elsewhere, to others. I felt the standard optimism of the age: the future will be better than the past. The promise of unfettered growth is that one day, everyone might share this optimism.

Yet one generation's optimism yields to another's unease. My nephews' peer group was the first to believe they may not have the same opportunities as their parents. My son's is likely the first to normalize a dreaded and unknown future. Fires, droughts, and storms fill the nightly news; one can only feel blasé to hear about yet another record heat wave, see more parched fields and failed crops, or watch people paddle along flooded streets in little boats. The wars of the Arab Spring are as much indicators of environmental stress[iii] as they are uprisings against tyrants. Apocalyptic visions permeate popular culture. The young react with humor, rolling their eyes: "Look, another sign of the apocalypse!" They sense something's amiss with our roaring global economic machine. They know endless growth powered by burning fossil fuels is a Faustian bargain.

We face a paradox: the very market forces that created so much wealth now bring levels of climate risk that threaten economic security, even our civic structures. Business as usual takes us well past the nominally safe level of 2°C of warming[iv] and into very hostile territory. It's hard to overstate that risk. Attempts to capture it sound extreme, even silly; life in the twenty-first century may well become the stuff of nightmares and dystopian films. It's not a stretch to say climate disruption is the endgame of industrial civilization.

In principle, this is a technical problem: our task is to rebuild global energy systems and transition to a low-carbon economy. It's perfectly doable. Sufficient capital sits in pension funds and money-market accounts. The engineering and entrepreneurial talent is fired up and ready to go. The same forces that built the largest energy infrastructure in history can do it again. We know a price on carbon unlocks the capital, motivates the talent, and fires up the industrial machinery.

Had we tackled this challenge when it was first identified back in the 1960s — or when alarm bells began to sound in the 1980s, or even as late as the 1990s, when leaders gathered for the first Earth Summit — small, unthreatening changes to our economic engine would have done the trick. Then, we could have mitigated climate risk without destabilizing the

---

iii   The Syrian war, for example, was partly ignited by record numbers of farmers moving to crowded cities as their farms failed because of drought.

iv   Based on our current levels of emissions, we're already committed to at least 1.5°C–2°C of warming, and we're on a path to 4°C–6°C.

economy or requiring a radical rethink of economic priorities. The main difficulty lay in it being a collective action problem: the changes were small enough to warrant doing the right thing and simultaneously seek long-term economic advantage by developing solutions, but no one was motivated to go it alone, since everyone benefited in the short-term by holding back.

It's no longer that simple. What was once a technical challenge with collective action overtones has become a profound and systemic problem for three related reasons.

First, we've been unable to act because market intervention is incompatible with the dominant free-market ideology defended by a select portion of the economic elite. It's no secret powerful lobby groups[v] distort the democratic process to block action and well-funded libertarian neoconservative think tanks suppress our will to act by sowing doubt about the problem.[vi] The U.S. is a basket case of corporate takeover of government,[vii] and America's inaction hinders the rest of us. Market intervention remains a profound threat to libertarian sensibilities. Reasoned argument might win over time, but time is what we don't have.

Second, because we're so late to act, only vertiginously steep cuts in emissions[viii] will do. The time for easy, incremental change is long gone. Since the late '80s — when Margaret Thatcher warned the UN of the

---

v   The irony, of course, is that big business using its capital to distort the political process to defend the status quo is not compatible with free-market ideology.

vi   See *Merchants of Doubt*, by Naomi Oreskes and Erik M. Conway, for a compelling account.

vii   Fossil fuel influence over state institutions in the U.S. hit a high-water mark in the Trump era, with the appointments of Rex Tillerson (ex-CEO of ExxonMobil) as secretary of state and Scott Pruitt (an avowed enemy of environmental oversight) as head of the U.S. Environmental Protection Agency (EPA).

viii   The aggressive pledges made in Paris at COP21, if they are all met, bring a limit of around 3.5°C. The cuts required to hit 1.5°C are more than 7 percent annually starting now, which has never been achieved in history, with a near-exception in the collapse of the former Soviet Union. Even 2°C means 4 to 5 percent annual cuts. See Baer, P., T. Athanasiou, and S. Kartha. "The Three Salient Global Mitigation Pathways Assessed in Light of the IPCC Carbon Budgets," *Stockholm Environment Institute* (2013). jstor.org/stable/resrep00380.

unprecedented threat to global stability posed by climate disruption — emissions have doubled, accelerating alongside global trade and the rise of developing markets. Since then, climate disruption evolved from an abstract risk to real threats on the ground. Incrementalism is no longer an option. Only radical, even warlike, effort can save the day.

Third, those necessary emission cuts are now steep enough to bring their own economic risk.[ix] The only time we've seen equivalent emission drops were in times of economic ruin: Russia's economic collapse in the 1990s and the recent Great Recession in North America and Europe. We face the perverse situation where the cure seems as bad as the disease. A moderate may be tempted to argue it's better to go slow, keep the economy strong, and face what storms may come. I believe this thinking mistaken, but conventional wisdom that favors incremental change runs deep.

In other words, market fundamentalists blocked climate action for so long that the economic risk now posed by the radical market intervention we need is seen to be on a par with climate risk itself. The irony is palpable (when I get past apoplectic).

Blame isn't limited to those who blocked climate action. Those very moderates conveniently ignored the problem — clearly there for all to see — well past the point at which it was incumbent upon them to speak up. As recently as three years ago, climate risk was not on the radar of the global business elites who hobnob annually in Davos. As Price Waterhouse Cooper wrote in its annual report in 2016: "regulation, skills, national debt, geopolitical uncertainty, and taxes topped CEOs' list of concerns about threats to business growth."[2] It's inexcusable for those who benefit from, and represent, global corporate interests to have been so absent from this discussion for the past two decades. A deep moral pit has been dug, and it will take some climbing for that community to redeem itself.

Modern capitalism is, then, a mixed blessing. It has undoubtedly been the driving force behind the largest accumulation of wealth in the most

---

ix   Stopping at 2°C requires leaving about four-fifths of all known fossil fuel reserves in the ground. These reserves sit on the balance sheets of energy companies. The assumption we'll burn them all drives their stock prices. The collapse of those share values implied by limiting warming is known as the "carbon bubble."

complex and technologically advanced civilization in human history. On the other hand, it may well sow the seeds of its own destruction in being systemically unable to respond to catastrophic climate disruption. This is the legitimate source of deep unease and stinging criticism from the left.

Social critics Naomi Klein and Pope Francis are correct in their condemnation of *unregulated* capitalism. Unchecked market forces will bake the planet, destroy the economy, and pass untold injustice to the poorest among us and to subsequent generations. And these forces have deformed our democratic institutions by allowing power to accumulate in the hands of an elite that has actively sabotaged climate action for more than three decades. As a result, Klein wants to throw capitalism itself under the bus and build a new kind of economy. Which is not only a profound mistake, but politically incoherent.

As a practical matter, the democratic uproar needed to build whatever alternative economy Klein and the Pope have in mind is far greater than the upswell of the Climate Capitalism I'm proposing, which harnesses financial markets in the climate fight. Reengaging our political system to reform financial institutions like the World Bank, motivate the quantitative analysts (quants) on Wall Street, and redirect trade agreements to accelerate climate solutions is faster and more effective than waiting for something akin to Che Guevara's *revolución*. I will admit I simply don't know what that revolution looks like. Nor how we manage a complex modern economy without market forces. Those who've tried (today's Venezuela comes to mind) failed miserably. And none of the far-left socialist experiments of the past gave up growth — the primary bugbear in Klein's view.

In a sense, there's no going back now. We can only reform what we've got. A firestorm of economic instability unleashed by a complete rejection of market dynamics will take decades to peter out, at which point it's too late for the planet. Again, if the target is libertarian notions of free markets, sure. But that's not the same thing as criticizing capitalism itself, an argument that verges on the vacuous or incoherent.

I may appear more dismissive than seems reasonable of those working to build a new kind of economy based on empathy, connection, and respect for nature. They have a lot to offer those who have to live through the social breakdown my kid will see. When bread is ten bucks a loaf, or Pakistan runs out of water, our social structures and shared values will determine how we react. Thinkers like Klein and Pope Francis shape social resilience and marginalize zero-sum thinkers like Trump. Hence, their value may be best applied to adaptation and social resilience, not mitigation.

The trillions of dollars that sit in money markets and pension funds is the most powerful tool in our climate arsenal — if it can be redirected. We need to co-opt capital markets, not slay them. That capital is conductor for the rest of the economic orchestra. With it, we unlock the financial, engineering, and entrepreneurial might that can rebuild global energy systems. To think otherwise is naïve — the supply chains are too complex, the scale of manufacturing and project development too big, and the degree of entrepreneurial innovation required too deep. Like it or not, we must harness the very market forces that threaten our planet, to save the planet.

That said, Klein and other critics on the left are often right on target. Tepid market interventions that prioritize ease of transition over speed, cater to protecting vested interests, and avoid squaring off against a powerful economic elite that remains opposed to radical changes in economic priorities are a distraction. Cuts to emissions must be steep and swift. Market intervention must be correspondingly stiff. Economic casualties will result as high-carbon sectors are forced to contract, assets are stranded, the cost of high-carbon fuels goes up, and warming begins to bite. And those who invested in those sectors must bear the cost of making the wrong decision; we cannot afford to offload those liabilities to the public purse. Adaptation will be expensive, and that's where we need to spend scarce public resources. Adaptation means sensitivity to those affected and new protective infrastructure.

The Davos elite were read the riot act by a wonderful new interloper. In 2019, they were lectured to great effect by a brave sixteen-year-old Swedish girl, Greta Thunberg, who calmly argued, "Some people, some companies, some decision-makers in particular have known exactly what priceless values they have been sacrificing to continue making unimaginable amounts of money. And I think many of you here today belong to that group of people." After an awkward pause, she got some applause.

Maybe the billionaire class will step up, maybe not. But this problem is too intractable for billionaire heroes to play anything but a supporting role. An empowered public sector, backed by a strong democratic mandate to act, provides the framework in which markets operate. Our collective self-interest is expressed when that framework guides the economy toward a low-carbon state. This is a job for democracy, not billionaires.

Location seems to matter. Critics of climate action in Canada are fond of pointing out that we are responsible for less than 2 percent of global emissions. So why act? Reducing our emissions won't have any impact on warming levels, they say. It's true that no country, company, energy project, or individual can solve the climate problem going solo. But it doesn't follow that we, or anyone else, shouldn't act.

From a narrowly selfish perspective, economic advantage is conferred on those who get an early start. In a market economy, if you solve a big problem, you get a big reward. The world will de-carbon, at some pace and in some way, whatever Canada does. It's already happening, creating an enormous market. Countries and companies who gain a lead in delivering low-carbon solutions will face massive demand as the world moves away from high-carbon energy. Instead of exporting carbon-heavy oil, Canada can export innovative cleantech. Getting just our pro rata share of the global market (what one expects just for showing up!) would make our cleantech sector larger than our auto sector by 2030. That's what's at stake for those moved entirely by self-interest.

And there's more to life than narrow self-interest, of course. It's churlish to defend complacency by arguing that no single act makes the definitive difference. By the same reasoning, no one need vote. Humanity doesn't accomplish great things by always asking "what's in it for me?" The Canada I know steps up to the plate because it's the *right thing to do*. We have a long history of being on the right side of the great moral questions of the time. From jumping into World War II, to blue-helmeted peacekeepers, to protecting escaped slaves and Vietnam draft-dodgers, good citizenship means taking a moral stance. Climate is the single greatest moral challenge of this century. Acting aggressively today is the right thing to do, just as taking a position against slavery was over a century ago.

There are lots of places markets shouldn't go. In every developed country outside the U.S., some portion of health care is placed directly in the public sphere. Likewise, shared infrastructure, like roads and sewers, falls squarely onto shared shoulders. There is no climate mitigation without massive public investment in public transit. We will not alleviate transportation emissions (never mind congestion!) without fast, effective, and low-carbon mass transit. That includes subways and raised rail to serve dense urban centers, as well as high-speed rail to connect those cities.

And that can be effectively built only by governments. Partly because it's shared physical infrastructure, like a sewer, over which a monopoly can be more efficient. But, more importantly, this is because there are so many indirect benefits. London, like most modern cities, cannot operate without legions of low-paid workers, and those workers cannot afford to live or drive downtown. Want a barista to serve you coffee? A teacher to show up to your daughter's school? Nurses in the local hospital? Public transit is the lubricant without which modern cities would seize up. Indeed, I'd argue many profit-seeking rideshare enterprises (like Uber, or Lyft) sometimes make that job more difficult, not less. They siphon off the more immediately profitable routes, rendering the rest a transit ghetto, harder and more expensive to service. Market forces are important, but they are not the only game in town.

This book has three major sections. Part One, "What's the Problem?" outlines the major issues that need to be tackled. It starts in Chapter One, "The Debate: From Popes to Apologists," by describing the wide variety of responses to the climate crisis. The far left is led by Naomi Klein. The far right includes climate quietists like the Fraser Institute. Beliefs in the capacity of technology to replace fossil fuels vary widely. Techno-optimists like Elon Musk get on with the job while energy analysts like Vaclav Smil exude deep pessimism. Each position has merit and a contribution to make. But none has *the* solution. A common-sense approach takes the best of each view, applied to a piece of the problem, and jettisons the rest. From across the spectrum emerges a patchwork of ideas suited to the complexity of the problem.

In chapter Two, we explore "Capitalism: Caveats and Critics," the implication that capitalism means different things to different people.

For some, it's opening a small independent business. To others, it's unfettered flows of global capital. It's one thing to a neocon and another to a climate hawk. Canada, Sweden, and Russia are all different kinds of capitalist countries. Capitalism is not incompatible with a forceful public sector, the distorting presence of oligarchs, or local co-ops. In this view, throwing out capitalism, understood more broadly, makes no sense whatsoever. By better understanding its full scope of possibilities, we can better imagine what Climate Capitalism might look like.

Part Two, "Climate Capitalism: Carbon, Politics, and Solutions," gets to the nuts and bolts of building an effective Climate Capitalism as a blueprint for action. Chapter Three, "Price and the Need for Speed," argues that the efficiency of carbon pricing as a primary climate response is firmly established as economic orthodoxy on both the left and the right. Hence, current political controversies over it are but cheap political opportunism. Pricing carbon is the foundation of Climate Capitalism — but it's not enough. When your house is on fire — and ours most certainly is! — you care more about the amount of water the hose delivers than you do about its efficiency. With the speed of the remedy in mind as a clear priority, we can better judge efforts to accelerate a climate response beyond that basic foundation.

Chapter Four, "White Hats and Black Hats," argues that corporate and investment leaders must be partners in the transition to a low-carbon economy. A few — the White Hats — are willing to lead the way. But we won't be saved by enlightened executives; they're far too few. Some — the Black Hats — are overtly obstructionist. Thus far, these nastier types have been effective in stalling action. But it's not as simple as good guys and bad guys. The vast majority — the middling Grays — are decent people doing their jobs as best they can. Most have good intentions (Black Hats aside) and follow the rules. Our central challenge is laid bare: if corporate executives are mostly decent people following the rules, then those rules must be changed so that the full force of global capital and corporate might is deployed — now — in the climate fight. And only the public sector can make and enforce those rules.

Chapter Five, "Filling the Gaps," gets into the details of how we might build the infrastructure of an effective Climate Capitalism. Cleantech faces unique hurdles, especially given how quickly we need to accelerate adoption. The capital-intensive nature of energy projects means the technology isn't driven by consumer demand the way iPhones are. Cleantech is big wires, not little ones. It's giant machines as much as a distributed, smart energy grid in your home. Cleantech is not so much a sector of the economy, but a new way of doing business — affecting every industry from forestry to mining to furniture. A hodgepodge of solutions is layered on top of a carbon price for more comprehensive and effective climate action: green banks, flexible regulation, financial innovation, retooled trade deals, and better data on risk management for the corporate sector.

Finally, Chapter Six, "How Sovereign Is Sovereign?," looks at how Climate Capitalism plays out within increasing levels of sovereignty — from provinces and states to countries, and to the ultimate sovereign: Earth itself. Nations' internecine squabbles over their respective share of the burden to act often reflect the way that nation states have for so long pointed fingers and shirked responsibility. Perhaps as Canada learns to share responsibility between federal and provincial governments, we can provide a model for multi-jurisdictional climate governance. But global climate compliance is better served with enforcement mechanisms, which already wait in the wings to arbitrate when the inevitable bickering starts. There's no reason existing institutions — like the World Trade Organization (WTO) — can't be marshaled into service as the earth's economic policeman.

We can no longer avoid the Anthropocene, an era defined by an earth system irreversibly changed by industrial activity, but we might avert the worst of it (if we work very hard). The final section, Part Three, "Welcome to the Anthropocene," looks at two ways this century might play out. In one, Climate Capitalism flourishes. In the other, we just muddle through. Nobody knows in detail how unchecked warming will play out, but we do know it won't be pleasant. By facing the risks head-on — economic, social, even existential — we might muster what it takes to commit to Climate Capitalism. I debunk the idea that we can painlessly adapt to this changed world, or that adaptation is a better deal than mitigation.

In the battle of ideas, unregulated capitalism is on the wrong side of history. Wall Street took care of that by delivering unprecedented financial instability[x] in the 2000s and, more broadly since the 1980s, levels of inequality not seen since the Roaring Twenties. A renewed Climate Capitalism — one that aggressively targets emissions, faces up to market failures, and addresses the economic imbalances that result — can regain lost legitimacy. It's in the best long-term interests of the economic elite to publicly and aggressively fight for it.

In hindsight, the causes of the French Revolution are obvious. A starving populace got angry with an elite that had lost any sense of *noblesse oblige*. Climate catastrophe may bring no less a confrontation between a scared populace and an elite seen as out of touch, even culpable. That potentially violent cultural divide can't be softened by a last-minute show of concern. Indeed, elites are already beginning to retreat further into gated communities. Some build private compounds far from the madding crowds in extreme efforts at self-sufficiency.[xi] Aside from stoking resentment, these isolationist moves are absurd. None of us can jump ship. We really are all in this together.

In the distant future, the record of our time will be found primarily in the spike in atmospheric carbon traces we leave behind. Our geological record is already set. We will be known as the Carbon People. Whether or not those distant relatives look sympathetically on their deep descendants, only time will tell. My fear is not an economic revolution, but the lack of one.

---

x   The socialized costs and private gains of the recent financial crisis showed *unregulated* capitalism isn't all that different from *crony* capitalism, where capital is allowed to gain a grip on the political system.

xi Nigeria's offshore, ocean-walled Eko Atlantic city is a striking example of how the superrich will try to seal themselves off from climate impacts. See Lukacs, M. "New Privatized African City Heralds Climate Apartheid" *The Guardian* (Jan 21, 2014) theguardian.com/environment/true-north/2014/jan/21/new-privatized-african-city-heralds-climate-apartheid.

PART ONE

WHAT'S
the
PROBLEM?

"We certainly don't have time for wholesale changes in our economic system, because that sort of thing uses up all the available political energy for decades; if you want to overthrow capitalism, leave it for later."[3]

— GWYNNE DYER

CHAPTER ONE

THE DEBATE –
FROM POPES TO APOLOGISTS

Is modern capitalism capable of solving the unfolding climate crisis? Is a stable climate compatible with the voracious growth demanded by the modern capitalist global economy? The answers aren't obvious.

At one extreme, social critics like Naomi Klein and Pope Francis argue that unchecked capitalism and market forces are the *cause* of the climate crisis. Only by chucking out our growth-addicted, free-market economy can we meaningfully curb global greenhouse gas emissions. Klein's bestseller *This Changes Everything* and Pope Francis's second encyclical, *Laudato si'*, set the standard for the anti-market crowd. At the other end of the spectrum, uber-entrepreneurs believe those same market forces are our salvation. Only capitalist markets are capable of creating the innovative clean energy technologies and massive deployment we need. Led by the likes of Bill Gates, George Soros, and Jeff Bezos, they're kick-starting the effort by seeding the Breakthrough Energy Coalition with a few billion dollars.

Lots of views sit somewhere between. "Techno-optimists" believe innovation will naturally bubble up from the bottom and disrupt fossil fuels the way Uber did the taxi industry. Many environmentalists argue we just need to get the big, bad energy companies out of the way so existing solutions like solar and wind can blossom. Nuclear is particularly controversial. To advocates, it's the only zero-emission energy source that can play in the big leagues. To opponents, it's yesterday's dangerous white elephant. Some particularly jaded folks have given up on climate mitigation altogether. According to them, our addiction to fossil fuels can't be broken quickly enough, so we'd better get on with planetary Band-Aids like sucking carbon out of the air or geoengineering to cool the planet.

Fossil fuel apologists like Bjørn Lomborg and the Fraser Institute flip the climate issue on its head and frame increased emissions as a moral issue wrapped in free-market garb. The developed world's wealth was built on fossil fuel, the story goes, so we should let everyone else burn lots of the stuff so they can get rich, too. Limiting carbon emissions is like taking food and schooling from the poor and hungry masses! And being wealthy is the best defense against a changing climate. So, better to keep the coal fires lit and batten down the hatches than try to avoid the storm: adaptation good, mitigation bad. There's nothing like a love of fossil fuels to bring out the far right's professed love of the poor and needy.

And there's always the question of cost. There's no free lunch in a capitalist world, and solving this problem looks expensive. Climate action may be the right thing to do, but it's a drag on economic growth. Or is it? I (and many others) argue we're better off in the long run switching from resource-based energy sources like natural gas, oil, and coal and transitioning to technology-driven sources like solar, wind, and storage. But even if that's true, a radical rewiring of our energy systems comes with real up-front costs. Solar might be free once it's up and running, but someone has to buy the panels and the storage facilities to make sure the lights stay on after the sun goes down. Who pays?

Developed countries built our roaring economies on a foundation of fossil fuel. Does that mean we owe an historical debt to act first, to help less-developed countries get off fossil fuels and adapt to a changing climate? When the Maldives disappear under a rising ocean, can they sue us? International climate negotiations have bogged down on these issues for decades. Current behind-the-scenes negotiations in North

America indicate some kind of trade-off whereby the fossil fuel majors get indemnified from climate damages in exchange for their support of a carbon price. Is that fair? What price, and for what degree of indemnity?

It's true that historically, developed countries emitted the majority of emissions. That speaks to a responsibility to take the lead on climate, but it's not that simple. The resulting modern economy produced a technological base from which all developing countries benefit. They don't have the same development cycle that, say, Britain or the United States went through. Hence, they inherit an ability to short circuit most of the historical development trajectory. Those benefits are direct and substantial. They can leverage low-cost modern energy technology as they seek to mitigate their own emissions and even gain economic advantage. China did not invent, nor commercially develop, the solar panels they now sell to the rest of the world. While there is a moral argument that developed nations must take the lead, it's not as clear-cut as it might appear.

Regardless, the Canada I know doesn't shy away from punching above our weight on the great ethical issues of the day. Our diplomatic strengths shouldn't be underestimated. We led the multinational framework that did away with ozone-damaging chlorofluorocarbons (CFCs), for example. We are well-suited to show leadership on climate change. Resolving our provincial-federal squabbles may serve as a model for intranational disputes — how to fairly and effectively distribute emission reduction responsibilities among subparts of a larger economy? And dealing responsibly with the heavy-oil assets we have in the ground (leaving most of them there) would go a long way to convince other petro-states to do the same. Bottom line: if a rich, comfortable country like Canada can't make a good-faith effort to limit emissions, we have no business asking others to do so. That's not my Canada.

Somewhere in the middle of this raging debate, reasonable-sounding economists quietly urge a variation on wartime British resolve: "Keep calm and price carbon." And if it's made revenue neutral by lowering other taxes an equivalent amount, we might secure the endorsement of a right wing opposed to all things tax-like. They don't even need to care about climate! With this view, no one gets hurt. Everybody wins. We break our addiction to fossil fuels gradually — carbon price as methadone. Yet the urgency, scale, and pace of change we need belies such a soft landing. Surely it can't be that easy? Spoiler alert: it's not!

The many views on climate reflect different backgrounds, priorities, and assumptions. Some have more merit than others. A few are ill-informed or disingenuous — the result of willful blindness, propaganda, or ignorance. But in each there is at least a grain of truth, some core idea or motivation that's intuitively valid. The cacophony of voices reflects the hard truth that weaning our economy off fossil fuels is the most complex and difficult problem humanity has ever faced. It makes putting a man on the moon look like a walk in the park.

As times get tough — and they will — people look for villains. Populist politicians find scapegoats. Deserved or not, the business community will make a good candidate for climate villain; Klein and the Pope are winning the battle for mind space on that front. On the other hand, for too long, the loudest corporate voices have been those who hijack action in this sphere. It's time for others to speak up, to acknowledge the trouble we're in, and endorse a difficult economic transition. It's enough to say, "We don't have all the solutions. But we get the problem, and we're going to try." The alternative is being on the wrong side of history — the villains of the story. In which case, we have the ugly prospect of decades of combative protests and increasing odds of real populist revolution at the ballot box that may well upend the economic system we know (mostly) works.

Perhaps most difficult of all: in an age when mistrust of elites is normal, climate risk asks that we place our trust in experts. Those who spend their professional lives studying climate science have increasingly bad news. As the scientific consensus and degree of certainty grows, so, too, has the level of alarm. It's always tempting to ignore bad news, especially when it's countered by the comforting noises of a slick, well-funded campaign of disinformation. But the atmosphere cares nothing for our cognitive comfort. And nature always bats last.

PARIS PROMISES AND THE PACE OF CHANGE

One of the first foreign visits for a freshly elected Prime Minister Justin Trudeau was to attend the twenty-first annual UN Conference of the Parties (COP21) in Paris in the fall of 2015. Trudeau was flanked by provincial leaders as he gave his opening address to the delegates. "Canada is back, my friends. We're here to help," declared the newly minted federal leader. His timing could not have been more fortuitous. COP21 had a markedly different mood from previous gatherings right from the start. And for the first time in years, the lengthy gabfest resulted in what was hailed by many as a meaningful agreement. Declared global ambitions were high. Action was imminent. The optimism of our "sunny ways" new prime minister seemed in tune with global sentiment.

It was quite a turnaround, both for Canada and the COP process itself. At previous conferences, Canada, under Stephen Harper's leadership, was rightfully painted an obstructionist. The number of "fossil of the year" awards Canada won rivaled our hockey golds. This time was different, and Team Climate Canada included some hefty support from Catherine McKenna, our negotiation-savvy minister of environment and climate change. Tapped by French foreign minister Laurent Fabius, McKenna helped shepherd more than a dozen countries across the finish line in a dramatic last-minute push. The world upped its stated ambition, formally endorsing not only a hard 2°C limit to warming, but vowing to make efforts to stop at 1.5°C.

In one year, Canada went from obstructionist to enabler. And global leaders went from bafflegab to speaking of meaningful targets. Triumphalist quotes flooded social media:[4] French President François Hollande told the assembled delegates, "You've done it, reached an ambitious agreement, a binding . . . universal agreement. Never will I be able to express more gratitude . . . You can be proud to stand before your children and grandchildren"; UN Secretary Ban Ki-Moon, called the agreement "a monumental triumph for people and our planet"; French foreign minister Laurent Fabius said, "[The delegates] can go home with their heads held high . . . Our responsibility to history is immense"; Christiana Figueres, then-executive secretary of the UN Framework Convention on Climate Change (UNFCCC) and a key architect of the

COP21 agreement, said "One planet, one chance to get it right, and we did it in Paris. We have made history together."

Wow. Sounds like good news, right? The planet's health in capable hands? Well, maybe. It's one thing for a kid to declare she's going to play on an Olympic hockey team and win a gold medal and quite another to actually make the team (never mind win the medal), even if her parents, coaches, and teammates are all on board. That's a bit like COP21: declaring an intention to stop at 2°C is akin to making the Olympic team, and 1.5°C is winning the gold. The targets are meaningful only insofar as they express admirable degrees of ambition. How likely are those ambitions to be met?

Not very, unfortunately. There are a few ways to look at how we might meet any given target: how soon we need to drop net emissions to zero[i]; how much carbon we can emit in the meantime (the carbon budget); and the sheer scale of the zero-carbon energy infrastructure we need to build to replace existing emissions and accommodate growing energy demands.

1.5°C: A Quixotic Dream?

Let's start with the more ambitious goal of limiting warming to 1.5°C above preindustrial levels.[ii] Despite the hopeful talk, that's not going to

---

i   No matter the long-term temperature target, we need to get to net zero emissions. Only a net emissions value of zero provides a stable level of atmospheric greenhouse gases (GHGs), since any net flow of emissions increases the total atmospheric stock. As that stock grows, so does the global average temperature. The lower the target, the earlier we need net zero: 1.5°C within a few years and 2°C well before the end of the century.

ii   Levels of warming can be confusing because there are different starting points (or baselines). Some data sets (such as NASA's Goddard Institute for Space Studies) use the temperature difference from a baseline that's the average of 1951 to 1980. Others (like the National Oceanic and Atmospheric Administration, or NOAA) refer to anomalies from a twentieth century moving average. Since both of those baselines contain some degree of man-made warming already, we adjust data to refer to the more useful baseline of preindustrial temperature. A stand-in for that is the 1881–1910 average, since that's the earliest date for which global temperature data are reliable.

happen. Why? Because we're pretty much there already![5] And no matter what we do, there's more heat already baked into the near future. We've seen warming accelerate over the past few years as oceans burped out some of the heat they'd been storing temporarily. Critics pointed to a recent flattening of the atmospheric warming curve (since the late 1990s) as evidence we needn't worry so much. But the oceans are the great thermal battery in the atmospheric system, storing more than 90 percent of incoming heat. And they have kept heating up. Their inexorable rise was an easy clue as to what happened next.

Sure enough, even as COP21 was taking place, global temperatures spiked massively. In the three months immediately after the conference, we bumped against the 1.5°C ceiling; January to March of 2016 averaged 1.48°C, with February hitting 1.55°C. Yes, that was a temporary peak driven by a strong El Niño[iii] and is within a range of short-term fluctuation, and it's true that three months do not mark a long-term trend, but it's a strong indicator we're already committed to blow through the 1.5°C target. The longer-term trend agrees with that directional spike in warming: 2016 marked the third record year in a row, at 1.2°C above preindustrial times.

Ok, let's say we try. Really hard. What do the emission numbers look like on the 1.5°C target? We'd need to reduce global greenhouse gases by nearly 10 percent per year, every year. I don't know anyone, anywhere, who thinks those kinds of reductions are remotely possible absent a catastrophic economic shutdown. Even getting to net zero emissions tomorrow — an impossibility — means warming will still continue for decades beyond that because of the long lag it takes for the planet's warming to catch up to what's already out there in the atmosphere. Sadly, it looks like a 1.5°C rise is already baked into the system. When I hear people talking hopefully about that goal, I'm reminded of Don Quixote, who tilted at windmills while imagining himself to be living the heroic life of a chivalrous knight. It's a nice story to make us feel good, but it has little basis in fact.

---

iii   A regular cycle whereby the Pacific Ocean unloads large amounts of heat.

An Intergovernmental Panel on Climate Change (IPCC) special report on the target was released in 2018. It made for interesting reading. It confirmed that "limiting warming to 1.5°C is possible within the laws of chemistry and physics," but to do so requires the pragmatically unattainable global reductions of "45 percent from 2010 levels by 2030, reaching 'net zero' around 2050." That's ten years to reduce emissions by *half*! Unsurprisingly, it emphasized the need for late-century carbon suction[iv] and geoengineering[v] to square the circle.

I expect going forward to see a sleight of hand in public discussion of targets to save face while politicians quietly shift baselines to something else, like a comparison of the modern average temperatures to 1985–2005. That gives 0.6°C of breathing room; but it's just an accounting trick, like hiding debt by moving it off the balance sheet. Or, in keeping with our hockey analogy, settling for the local triple-A hockey team and pretending that's what you meant by "Olympics" the whole time.

2°C: A Stretch Goal

The more moderate 2°C target remains *barely* possible. It's what one might call a "stretch goal" and may be achievable with an immediate, comprehensive switch to the Climate Capitalism outlined in this book. Recent analysis puts the odds of us reaching the 2°C targets at a one in twenty[6] chance. We'd need to get to net zero emissions somewhere in

---

iv   There are three options: sucking carbon dioxide directly out of the atmosphere using energy-hogging technology (like that being developed by Gates-backed Carbon Solutions), planting vast numbers of trees (unlikely to improve things since the die-off of rain forests is expected to more than counter the effect), or carbon-negative electricity plants that burn biomass and capture and sequester the emissions (Bio-energy with Carbon Capture and Storage, BECCS). More on this later.

v   Direct manipulation of the global temperature by, for example, putting dust in the upper atmosphere to deflect sunlight. More on this later.

the middle of this century,[vi] but most importantly, we must start reducing now — and get a 20 percent reduction globally by 2030.

There are an infinite number of ways in which that work can be distributed. The main divide is between the goals of fully industrialized countries with mature economies and energy systems and developing countries with fast-growing economies but immature energy systems. The differentiated responsibility between the two will be an ongoing dispute, but one can approximate a scenario.

Following the basic carbon stock and flow models provided by Climate Interactive, if we focus just on the short-term — a plan that covers now to 2030 — we can put some flesh on the bones. Hitting a 2°C target means dropping emissions in the developed countries (the EU and North America mainly) by half. China and the rest of the developing world could peak by 2025, then drop by 2 percent a year. That's tough, but doable. The main hurdle to early and deep emission cuts is the expected increase in energy demand due to increased wealth and population in developing countries, meaning they can't meet the same targets as North America and the EU. But it's also what makes them less predictable; since their energy systems are immature, and technology is changing so fast, we have no idea what's possible there.

The good news is clean, renewable energy is already the dominant new energy source, particularly in the developing world, where electrical grids remain nascent. Even oil-soaked countries like Saudi Arabia are turning to solar in an effort to turn down the spigot. The bad news is that our total energy usage, from sources old and new, is still far too heavily weighted to fossil fuels; and the developing world has a long way to go in economic development before they come close to the developed world's standard of living, which means using lots more energy of some kind to get there. And there's no way I can see they'll be denied the chance to catch up. This is what makes the work so tough — we expect to *double* total current global energy output by 2040.

---

vi   Just to throw in some further pragmatic constraints: if we limit total spending on mitigation to 3 percent of GDP and disallow future atmospheric carbon capture from the accounting, for example, we would actually need to get to zero emissions by 2030.

The wild card is technology — its effect on all systems, including energy, is often fiercely underestimated.[vii] But the scale of deployment is daunting: to hit 2°C, half our energy needs to come from renewables by 2028.[7] In 2014, it was just under 10 percent. As an example, we'd need to ramp up wind installation rates by nearly forty times in only fifteen years. Possible, but only if we put our industrial economy on a warlike footing, as the United States did following the attack on Pearl Harbor. That takes a degree of political will we don't yet see.

In any case, the 2°C target requires a drop in emissions roughly equivalent to what we saw in Russia when the Soviet Union collapsed. Or briefly approximated in North America when the Great Recession hit. But instead of a couple of years, we'd need to do it for decades. Economic collapse is no guide to action, of course, but it gives a hint of just how pervasive emissions are to economic activity, and how difficult it will be to grow the economy while simultaneously lowering absolute emission levels by substantive amounts.

Places we've seen the most success in proactive, economically positive reductions were either insufficiently shallow or lasted a brief time. British Columbia, for example, had a nice drop in emissions when their carbon tax came into effect, but it was short-lived as the easy-to-do stuff ran out of steam and the economy continued to grow. The United Kingdom maintained a longer track record, but the levels of reduction are mild compared to what's needed. There is little historical precedent to provide comfort to those who believe we can tinker around the edges of our economic machine and hit the stretch goal.

The takeaway is a difficult but central idea upon which Climate Capitalism rests: a reduction in developed world emissions by half in fifteen years will not happen under business as usual, or moderate incremental changes in tax policy. It's too late for that. Radical changes are

---

vii   Reputable pinstripe groups like the International Energy Agency (IEA) routinely and massively underestimate the growth of renewables and their decline in cost. The only group close to predicting solar energy's dramatic rise in recent years was that woolly headed group of idealists at Greenpeace.

required even to preserve the status quo, best demarcated by that 2°C limit in warming.

3.5°C: Our Mutual Promise

The most realistic picture can be found in the aggregate pledges of individual countries. Almost two hundred countries brought promises to Paris, snappily named intended nationally determined commitments (INDCs). If everyone keeps them, we'd see about 3.5°C of warming.[8] Those pledges are not legally binding but are made on a "best efforts" basis. The only thing binding about the COP21 agreement is a commitment to be transparent about our emissions. We've agreed to keep an eye on each other as we all make an effort to meet our end of the bargain. Everything else is voluntary.[viii]

That's what makes Trump's administration so dangerous (well, one thing that does). The whole system for mitigating climate disruption is predicated on countries walking in lockstep, citizens holding governments' feet to the fire, and each country keeping a close eye on others to ensure they're doing their bit. As the U.S. breaks ranks, particularly as it did with a belligerent and bellicose denial of the basic facts of climate risk, there is every reason to think other countries will renege on their efforts. Why should India make an effort when the richest country in the world rudely abdicates responsibility? The long-standing problem of collective action was met with binding transparency and the trust of a promise. Lose trust, lose collective action. Trump, and a compliant GOP, may well bring down the whole house of cards. Mitigating that risk is the collection of cities and states within the U.S. that upped their game after Trump abandoned[ix] the Paris accord.

The good news is the economics of clean energy is already overtaking

---

viii   That's likely the only way it could ever work, given the straitjacket requirement of unanimous approval, which hobbled negotiations for the last twenty years.

ix   He can't officially abandon the accord, as the time period of notice to withdraw extends just past the next election date. But his vow to do so is enough to damage momentum in the way described.

the politics. Corporations, financiers, and energy companies around the world see the writing on the wall and are jostling for advantage in the coming low-carbon transition. Clean energy now is cheaper than its fossil fuel counterparts in many regions and sectors. Electric cars are beginning to make the internal combustion engine look old-fashioned. But it would be unwise to discount the dangers of a reemergent right wing in the United States that is (or professes to be) illiterate on the science, and whose appetite to feed on the short-term gains of fossil fuel extraction seems unbounded. A kleptocracy that profits by burning what coal and oil it can, while it can, deals the planet a lousy card at a critical time.

Absent a collapse of collective pledges, there's still nothing to celebrate about the politically palatable but environmentally disastrous implicit goal of 3.5°C. At that level of warming, Greenland will be gone, as well as large portions of Antarctica. Florida underwater. The U.S. Midwest a desert. Pakistan parched. Australia and California on fire. All coral reefs gone. Chaos. Yet those promises are likely the maximum that was politically feasible for each country at the time they were made.

Canada, for example, kept to the pledge made by the Harper government of a 30 percent reduction by 2030 even though Trudeau and McKenna would have liked nothing more than to righteously ramp up ambition. Why didn't they? In the cold light of real policy decisions — and early pushback from provinces, people, and fossil fuel profit makers — even that target looks ambitious. Canada is nowhere near on track, even with Trudeau's carbon tax combined with parallel provincial efforts including Alberta, Quebec, BC, and Ontario (prior to the Doug Ford government). We can get there, but it will take a lot more arm-twisting, creative economic policy, and a hard look at the potential growth of our most carbon-intensive industries.

Real Hope: Getting to a Positive Cycle

Clearly, we need to go further than current political dynamics allow. Real hope comes in the five-year review periods baked into the Paris Agreement. Periodically, countries are asked to renew or increase their pledges. There are three ways the door might open to greater ambition to cut emissions:

it's likely to be easier than people thought (at least the early stuff), particularly in the developing world; countries that gain economic advantage as early movers on developing solutions, with jobs to show for it, will be examples for others to follow; and, as environmental and economic harm begins to bite, and the downside risk becomes more clear, populations and business leaders will demand more action in their own self-interest. A cycle of positive reinforcement might kick in as we ratchet up effort commensurate with a changing economic and social environment.

But whatever targets we consider, we're up a creek. They are either impossible yet safe (1.5°C), near impossible and maybe safe (2°C), or difficult and dangerously high (3°C plus). Targets appear either insufficient or too hard to reach. The lesson is not to give up; the number of unknowns and uncertainties that permeate all of these (highly imperfect) models of both human and planetary systems means there is always hope. Particularly if and when vested interests who currently block action — like Charles Koch, the Trump administration, and free-market think tanks — decide to jump on the bandwagon as they see their sources of profit and/or credibility disappear.

The takeaway is not to focus so much on the targets as ends in themselves. They may provide a "north star" guidance system, but only directionally and not as an absolute endpoint. My view is we start by asking the simple question: Just how hard can we work? What's the maximum strain our economies might withstand in making the transition to a low-carbon state? Put the pedal to the metal and see where we end up. It's an unsatisfactory stance for those who want a measured, quantitative cost-benefit analysis — but, frankly, it's the best bet we've got. And as the young new global climate activist Greta Thunberg points out, perhaps we'll find that panic is a better motivator than hope.

THE LAY OF THE LAND

The debate about climate and its relationship to the global economy is confusing. And for good reason. Climate disruption is not just an environmental issue. Energy permeates every nook and cranny of modern life, from how we heat our homes and get to work, to driving the largest industrial base in history. So mitigating climate disruption requires that we put the economy on a fast track to a low-carbon state, and that means decoupling economic activity from fossil fuels. Climate is an "everything" issue.

In saying "this changes everything," Klein is absolutely right: climate is a moral, social, cultural, political, and economic issue. It affects all aspects of human life, from cultural norms to intergenerational justice. It exacerbates inequality, deepening an already troubled relationship between rich and poor. It challenges global trade rules and basic assumptions about economic growth. Every interest group has its own take on it, its own position, and its own potential solutions. And generally, they don't agree on who's to blame, let alone how to move forward to fix it.

The various and divergent positions of the primary groups are summarized below, each one emphasizing some part of the huge, all-encompassing change being wrought by climate disruption. Individually, no one school of thought has it completely right, but there's something to learn from each. And eventually, taking all these positions into consideration, one can see a solution emerge from what often appears to be a collection of wholly irreconcilable differences.

Naomi Klein Is Right About Something . . .

Naomi Klein has always been a cogent voice of conscience. Her book *No Logo* correctly identified how corporate branding deeply affects our collective cognitive culture. *Disaster Capitalism* skewered those who choose to be naïve about (or willfully blind to) the lengths to which capital will go to seek and expand markets. Her clear-eyed writing on the morality of social justice is a necessary counterbalance in a world where unrestrained free-market ideology can (and often does) wreak havoc. And Klein's recent climate tome, *This Changes Everything: Capitalism vs. the Climate*, is one of the best books on climate risk I've read. Except for her conclusion, about which she is astonishingly naïve.

Given her previous work, it was perhaps inevitable her proposed solution to climate risk is to throw out capitalism,[x] kneecap the big corporations, and shrink the economy. Klein follows a long line of critics on the left for whom the only possible solution to *any* problem — from inequality to environmental degradation — comes from nothing less than a total economic revolution in which a morally superior system, purged of greed and bad actors, comes into being. No one problem can be isolated and solved on its own, since it's the system that's at fault. A desire for ideological purity trumps a pragmatic response in this case. People who work to compromise are dismissed as insufficiently committed to the cause or blind to the underlying causes — the "real problem."

If this sounds unfair, look no further than the 2016 U.S. election. On the ballot in the state of Washington, voters were asked if they supported a carbon tax. It was one of the first direct tests of voter appetite for climate action. Being revenue neutral — all capital raised was returned in the form of other tax cuts — it was deliberately designed to attract support from across the political spectrum. Klein urged environmentalists to work *against* the initiative. Why? First, since the revenue wasn't spent on sufficiently progressive social justice causes it was too "right-wing friendly."[9] Second, because the initiative was "hardly enough to jumpstart an urgent, sweeping phase-out of fossil fuels,"[10] it was insufficient. This all-or-nothing tactic led to the tragic irony of the Sierra Club and Koch Industries fighting together to defeat one of the only climate bright spots in an overwhelmingly bleak political cycle.

In her criticism of the Washington initiative, two central confusions of Klein's book come into focus. First, she's correct that Washington's tax is insufficient to get rid of fossil fuels entirely. But no single initiative can be enough, certainly no local initiative. Must all partial solutions be rejected on those grounds? If so, there can be no first step

---

x   As a bonus, this will simultaneously bring about a new age of social justice and prosperity and equality for all. Klein seems to view climate disruption as the ultimate catastrophe, from which we will finally drum up the democratic will to overthrow capitalism and all its ills. I'm sure the irony is not lost on her, nor readers of *Disaster Capitalism*, which identified ways in which catastrophes are often used to entrench and advance ideological positions.

on a long journey — only an end point. Second, in demanding a progressive-friendly climate policy in a democracy deeply divided across multiple lines, Klein ensures the impossibility of political consensus. Without consensus, a sufficient solution is impossible.

When I politely criticized her position on Twitter, her response was to point out that a more socially progressive initiative was on the ballot for 2018 — one she and the Sierra Club would fully support. Trashing the former initiative in favor of the latter, she claimed, is "sometimes how progress is made."[11] All or nothing. Perhaps. Except that initiative failed as well. Would the first have passed if Klein and Sierra had got behind it? We'll never know. But we do know the more ideologically pleasing version did not. So, Washington state has nothing. Thanks for the help.

Her criticism of British Columbia's carbon tax follows similar lines: "two-thirds of the tax cuts have ended up in corporate pockets, while carbon emissions have been rising in recent years."[12] Here, she conflates the effectiveness of the tool (the carbon price) with its mode of delivery (who gets the money). If you want deeper cuts, raise the price. Whether the money ends up in corporate, personal, or government pockets is a matter of political preference.

The best of *This Changes Everything* is its realistic view of the scale of the problem and the time we've got left to act. Klein is correct to say incrementalism is no longer an option, and "there is no way to get cuts in emissions steep or rapid enough to avoid . . . catastrophic scenarios without levels of government intervention that will never be acceptable to right-wing ideologues."[13] The worst is its easy dismissal of the difficult work of generating consensus: "[climate] challenges something that might be even more powerful than capitalism, and that is the fetish of centrism — of reasonableness, seriousness, splitting the difference, and generally not getting overly excited about anything."[14] With all due respect, many of us centrists are plenty excited. And not all climate hawks need lean left.

Right-wing, libertarian ideologues are one thing, but discounting the entire political spectrum between Klein and those ideologues is another. Linking climate action to a non-negotiable far-left agenda marginalizes many in the business community and voting public — two groups essential to accomplishing those steep cuts Klein advocates. Pushing the environmental and business communities to work against climate policy that doesn't fit a broader progressive agenda[xi] divides us when we need to come together.

The real work is getting others who see the world differently to act in concert toward a shared goal. And, yes, that means compromise. That's where the work lies. Yes, it means we're unlikely to get a perfect climate solution. That's life. Resting on a sense of moral superiority that keeps one above the fray, untainted by doubt, looks brave, but it's really playing it safe. And it's not so different from those on the other side of the climate debate who deride as hypocrites those climate activists who still fly, drive cars, or heat their homes. Striving to retain a kind of moral purity in a complex world is not an asset but a liability. And holding action hostage until you get everything else you want is arrogant.

A more subtle mistake goes to the heart of Climate Capitalism. Klein's rejection of centrism — the waving away of those of us trying to be "reasonable" — conflates economic radicalism with political radicalism. However, I propose that one can endorse extreme economic intervention as a necessary reaction to climate disruption without the additional commitment that the intervention be politically radical. Imagine a massive $500 per ton carbon tax offset entirely by corporate and personal income tax breaks; politically centrist, but economically radical. With a far better chance of happening than Klein's revolution. Why? Because it doesn't ask the right to give up their entire political worldview. They are welcome to come along — as they must be if we are to find a solution to the climate problem.

Klein demands nothing less than a centrally controlled, top-down economic system that prescribes solutions rather than letting the market discover them, and she points out that "this scale of economic planning

---

xi   Klein is open about linking climate to progressive activism: "I was propelled into a deeper engagement with [the science of global warming] partly because I realized it could be a catalyst for forms of social and economic justice in which I already believed." Naomi Klein, *This Changes Everything* (Toronto: Knopf Canada, 2014), 59.

and management is entirely outside the boundaries of our reigning ideology."[15] The usual response to this sort of discredited stuff suffices: the modern economy is complex enough that a central supply-side planning department is not remotely capable of its management. And carbon emissions are not like DDT — you can't just ban them and prescribe a replacement. Fossil fuels permeate the entire economy. Only market signals that similarly reach every pocket of decision-making — from the personal to the corporate, from consumer goods to project finance — can muster the creative destruction needed to do away with them.

In part two, I profile some of the corporations and entrepreneurs who are inventing and building the technological and financial innovations we need: energy storage, low-cost, highly scalable solar power, next-generation biofuels from non-food sources, green bonds that skip the big banks, high-risk big capital for new energy systems. They don't operate by dictate nor in isolation, but are embedded with, and dependent on, global supply chains and financial markets. The partners they need are many of the large international corporations Klein has little time for, but who bring scale, engineering depth, and manufacturing might. While these are people of character who care deeply about the world — and many gave up well-paying alternative careers — they need markets and profits in order to make change. Without capitalist machinery, they cannot build the clean energy technology giants we need.

My disagreement with Klein's position is a central tenet of Climate Capitalism: I believe that we're better off co-opting the system we've got than investing the effort to foment a revolution to overturn that system. Gwynne Dyer said, "We certainly don't have time for wholesale changes in our economic system, because that sort of thing uses up all the available political energy for decades; if you want to overthrow capitalism, leave it for later."[16] In that sense, Climate Capitalism is centrist, but it's by no means committed to incrementalism, nor does it shy away from the radical economic interventions we desperately need. It just places those interventions within the existing framework, rather than advocating we throw it all out. After all, it's much harder to build something new than to tear something down. We should be highly suspect of those who would rip apart our economic and social fabric.

Regardless of one's political preference, I simply can't imagine any scenario in which our global economic machine doesn't keep running in much the same way for another couple of decades. That's all the time we have to head off the worst. So stop trying to replace the machine. Instead, change the fuel!

A hint to the motivation behind Klein's position can be found in her demonization of competition: "the triumph of market logic, with its ethos of domination and fierce competition, is paralyzing all serious efforts to respond to climate change."[17] She misses a critical point: if the competition is set up to *solve* climate disruption, then those powerful forces she resents so much would switch to work on the side of angels. It was cutthroat global competition that brought the price of solar panels down 90 percent over the past few years (a price cut now available to communes and corporations alike). Companies that provide low-cost clean solutions dominate, those that don't, wither. Markets operate according to rules imposed by elected governments. Competition is not inherently evil. Nor, I would argue, is it possible to live without it.

Competition is not restricted to markets, as Klein admits only a few sentences later. "Cutthroat competition between nations has deadlocked UN climate negotiations for decades: rich countries dig in their heels . . . poorer countries declare they won't give up their right to pollute." It is no less than a reframing of all human interaction that Klein seeks. "For any of this to change, a worldview will need to rise to the fore that sees nature, other nations, and our own neighbors not as adversaries, but rather as partners in a grand project of mutual reinvention."[18]

Good luck with that. And I say it sincerely, as it sounds wonderful. I just don't think a wholesale change of the human condition will arrive in time to solve our immediate problem. In the meantime, we are to block climate action within the existing economic framework — including that promising carbon price in Washington state — in order to wait upon the arrival of a new dawn of the human spirit, driven by empathy and embodied in social, economic, and political interactions the world over. This seems to me a childishly hopeful delusion and a dangerous distraction that serves only to separate us when we need to come together.

Every economic and political system in history has been subject to the usual human foibles, including competition and domination. Venezuela is no less subject to those forces than China, the U.S., or Sweden. Anyone who's worked in a local co-op has seen infighting that's a microcosm of what happens in the largest state or corporate boardroom. There is no system in which humans behave perfectly, where competition, avarice, and greed are eliminated. We act out our humanity on many stages, including as economic actors. No one has a monopoly on utopian blueprints, the right or left, the local or global, the private or public. Whatever system Klein has in mind, and I cannot for the life of me find it explicitly articulated in *This Changes Everything*, it will be as subject to avarice as the one we have now.

That said, if we see Klein's attack as limited to purists on the right — neoconservative, libertarian-leaning ideologues whose prior commitment to free markets prevents them from endorsing climate action — then her criticisms are dead on the mark. Her claim that "the 'free' market simply cannot accomplish this task [an 8 to 10 percent annual reduction in emissions]"[19] is correct. Her worries that climate will bring about "a brutal crash, in which the most vulnerable will suffer most of all"[20] is also correct. Further, she's right in arguing "the actions that would give us the best chance of averting catastrophe . . . are extremely threatening to an elite minority that has a stranglehold over our economy, our political process, and most of our media outlets."[21] Ditch the neocon Wild West, predicated on dark money informing political choice, that currently underpins the American far right — agreed!

But it does not follow that the way forward is to ditch capitalism itself. That Klein identifies a "small elite" that's threatened should give readers pause. It certainly opens to Klein the option to limit her attack to free-market neoconservatism. On that front, she will find many allies, including myself. But she doesn't. She cannot let go of her prior intellectual commitment to a left-wing economic utopia. That we need an "entirely new economic model"[22] is a stale and tired demand. If you don't like the way the game is run, then change the rules. Refusing to play at all is not a credible position.

And none of her proposed solutions — including a publicly backed

"Marshall Plan for the earth"— can possibly work without market forces to guide innovation, bring costs down, enable capital markets to fund it or large corporate entities to provide the required manufacturing and engineering capacity. At bottom, we have no choice but to regulate and reform capitalism. We have no choice but to harness the very economic forces that are so powerful they can bake the planet.

Dreams of perfect economic systems are a dangerous distraction. We have maybe a decade or two to deal with climate before things spin out of control. To coddle a large proportion of those on the left with a jarringly naïve view creates a dangerous rift at just the time we need to work together. A far lower degree of political fervor than what's needed for Klein's revolution could fully co-opt capitalist forces. It could rewrite trade agreements, use the WTO to enforce compliance on emission reductions, leverage the World Bank to finance and enforce climate solutions instead of forcing neoconservative ideology on the developing world, and entice the trillions of dollars of wealth sitting on Wall Street and in the City (London's Wall Street) to back climate solutions.

The "Leap" or Green New Deal: A Plan to Solve Everything

What sort of policy might Klein like? In early 2015, I was asked to endorse an emerging call to action called the Leap Manifesto, a document enthusiastically supported by Klein. Like the recent U.S.-centric Green New Deal, Klein's type of all-or-nothing view on climate is rendered as explicit policy recommendations for Canada. Subtitled "A Call for a Canada Based on Caring for the Earth and One Another," it asks that we rewrite our political and economic narrative along communitarian lines. In so doing, we would in one fell swoop deal with a laundry list of social ills in addition to the climate crisis, including the detestable historic treatment of Indigenous Peoples in Canada, racial and gender inequality, the global refugee crisis, childcare, and so on. As much a loyalty test of leftist credentials as a serious attempt at climate policy, it was not a document I could support — and not because its goals are not worth pursuing. It has two failings, one of which is unforgivable.

From a tactical perspective, it's unsophisticated and has little new to offer. It does suggest that we price carbon, retrofit housing stock, support public transit, and foster local engagement on energy systems. All good suggestions, but they are lost in a grab bag of other demands for immigration reform, refugee rights, and support for the arts. Transforming energy systems is harder than setting up local co-ops to build solar farms. And the pushback against trade inherent in this policy means those solar farms may well cost a lot more than they do now. Conspicuously absent are links to groups most critical to an energy transformation: utilities, engineering firms, entrepreneurs, energy regulators, banks. It reads as harmlessly nice, but it's not something those most responsible for its execution could endorse. The lists of initiating signatories and initiating organizations confirm this.

But it's from a strategic perspective that the Leap does real harm. Getting the left to agree on climate action isn't difficult — we've been there for years. A bit ragtag at times, and politically hobbled by an annoying first-past-the-post electoral system, the left's support is a given. The challenge is to get centrists and the right to come to the table. Just when the central challenge is to *come together*, the Leap further divides us by alienating precisely those groups.

For the more paranoid on the right, climate action is a disguised attempt at hijacking the market in a broad effort to inject an intrusive state into their lives. The Leap does nothing but feed that paranoia. For moderates on the right, there appears to be no room for those who invent, engineer, and produce energy infrastructure; the entrepreneurs, utilities, banks, engineers, and large corporations are marginalized, at best. The people in those institutions don't just make those hated pipelines, they make batteries, solar panels, wind turbines, and high-voltage transmission capacity. For centrists, the absence of markets, utilities, and banks gives pause. How are we to organize this new energy system? Who runs it? Who builds it? Klein and her supporters may have faith that community power systems are sufficient to run our industrial economy, but centrists — for good reason — do not share that confidence.

In the United States, however, where political corruption runs much deeper than in Canada, there is a strong counterargument. The

Green New Deal can play a role there that the Leap Manifesto could never play in Canada. In the U.S. today, Congress can't pass policies that a clear majority[23] agree on, including ones that advocate for higher marginal tax rates, stricter gun laws, and better health care. The climate argument goes like this: trying to placate the political class that controls Washington hasn't worked for twenty years, so do the opposite. Build something that generates enough excitement to bulldoze through an unresponsive Washington. Build a coalition that includes social justice warriors that don't identify as environmentalists, and bring in people concerned with local employment, health care, housing, racism, and inequity. Use that political hammer to force through a common-sense economic plan that a corrupt Washington has long refused to acknowledge. Give enough people enough to get excited about and maybe Congress will be reformed.

On this view, the right's recalcitrance on climate is a permanent feature of a political landscape that includes a number of other radical social tensions that have been building for decades. That is certainly Trump's America. Those other tensions can be tapped to ignite a political firestorm that legitimately lays claim to a broad rewriting of the social contract. While Canada is certainly not perfect, the U.S. is unique in the degree of social tension available to be harnessed in this way. In that sense, I am more sympathetic to the Green New Deal in Trump's America than the Leap Manifesto in Canada. The Green New Deal has the merit of existing in, and being responsive to, a political environment in which the depth of political disillusionment and degree of confrontation is unique and likely unstable.

But absent corrupt politics and an angry center, the Leap Manifesto and Green New Deal are divisive. By linking climate action to a communitarian view of the world, both alienate anyone unwilling to move to the far left of the political spectrum. Without enough support to overcome the inevitable political backlash, both wither on the vine. Worse, they provide fuel to those on the right who will use such policies as evidence that climate action is nothing more than an excuse for an intrusive state to overreach and threaten their own deeply held notions of individualism. The irony for such a communitarian group is that it's not by reaffirming our own convictions that we will act on climate, but by reaffirming those of others.

Breakthrough and Ecomodernists

One problematic commitment of capitalist economies is the need for constant, never-ending growth.[xii] A long tradition in environmental writing, from Rachel Carson to Bill McKibben, sees limits to growth as a cornerstone of environmental stewardship. Whether it's the pollution we emit, the forests we clear, the open-pit mines we build — less is more. Respecting the environment means limiting our effect on it. That much seems incontestable. But is the link between growth and ecological damage really that simple? To limit climate risk, must we target economic growth itself? The folks at the Breakthrough Institute don't think so. They argue the opposite.

Released a decade ago, Michael Shellenberger and Ted Nordhaus's controversial book *Break Through: From the Death of Environmentalism to the Politics of Possibility* was a sustained attack on conventional environmentalism's strategy. Critical to broad support for climate action, it argued, are two ideas. First, people don't care about the environment until after they've achieved some degree of material wealth and security. The developing world will not concern itself with climate until after their economies have grown. Second, we are more likely to respond when we hear an appeal to possibilities, rather than limits. Generating the political will to act on climate means talking about new economic opportunities, not constraints on behavior. In other words, we need more economic activity (albeit, of the right kind), not less.

Break Through argues a few simple premises: asking people to forego development won't work; frightening people will not make them act; showing them nature and expecting them to care enough to take action is ineffective. It argues that raising standards of living is a better approach to environmental care. On this view, the environment is a "luxury" good.

---

xii   My own views on this issue are provided in some detail in the next chapter.

Break Through argues that modern environmentalism needs a positive story that puts us in control of our destiny and unlocks new opportunities. As a result, they have an unapologetic "pro-growth agenda"[24] as a practical response to "the politics of limits, which seeks to constrain human ambition."[25] Traditional environmentalists make a strategic error in positioning "global warming . . . as [a crisis] of *too much* rather than *too little* progress."[26] Too many environmental strategies "aim to *constrain* rather than *unleash* human activity."[27]

I came to the same conclusion in my last book, *Waking the Frog: Solutions to Our Climate Change Paralysis*, but from a different angle. Our cognitive systems are wired to accept beliefs that corroborate our worldview and reject those that do not. That worldview is largely unconscious, and it is patterned through a lifetime of cultural influence. Progress, economic growth, and boundless human innovation and potential are central to how we view ourselves. Since the climate threat is incompatible with those beliefs, we tend to minimize or even ignore it. But a new economy of clean energy abundance, driven by human ingenuity and underwritten by an economic stimulus that creates new jobs, speaks to the best in us. That's no excuse to shy away from articulating the risks we face, but without offering positive solutions at the same time, we'll remain paralyzed.

Break Through's approach is tactical. Of course, the ultimate goal is to limit carbon emissions. It's a question of the most pragmatic way to get it done. The self-described ecomodernists[xiii] who run the Breakthrough Institute continue the book's tradition, with an emphasis on technology and human innovation as liberating forces to turn the tide on ecological collapse. They re-enforce climate concern as a luxury good and put economic development and massive energy production at the fore: "Plentiful access to modern energy is an essential prerequisite for human development and for decoupling development from nature."[28] With unlimited energy, progress marches with no end in sight.

Ecomodernists emphasize *separating* humans from nature: "We reject . . . that human societies must harmonize with nature to avoid

---

xiii "An Ecomodernist Manifesto," coauthored by Shellenberger and Nordhaus, is easily found online.

economic and ecological collapse." As we liberate the environment from the economy, they say, we will be able to "spare it for explicit aesthetic and spiritual reasons." With enough technological firepower concentrated in our cities and farms, we can "re-green" and "re-wild" nature. This separation is driven not by a desire to move away from nature, nor from a hubris that sets us apart, but rather a desire to spare nature from the various ravages of economic activity. Nature here is to be valued on its own terms and for reasons unlinked to economic activity. In that restricted sense, ecomodernism is deeply spiritual, not so different from radical groups like Earth First!.

Renowned biologist E.O. Wilson shares a similar view: "In the digital age, with the fundamental economic incentives humans have to get the most from the least, the economy evolves almost automatically toward less material, less energy, and more efficiency. This trend, combined with a move toward alternative energy sources, creates the potential of reducing what is called our ecological footprint: the average amount of space required by each human being for his or her livelihood."[29] Wilson is similarly concerned with limiting the intrusion of economic activity on nature. His goal, called Half-Earth, is to leave half the planet free from human intrusion with an eye to preserving genetic diversity. Wilson represents the latest in a long line of naturalists who have argued nature has value far above and beyond the mere economic.

The ecomodernists' commitment to nearly unlimited amounts of energy and a desire to wall nature off from economic activity leads them to endorse only the most muscular forms of energy, those that take up as little physical space as possible. Shellenberger and Norahaus's "Ecomodernist Manifesto" endorses more efficient (and undefined) next-generation solar, but its dismissal of existing solar, wind, and energy storage means it comes off sounding like an advertisement for nuclear. The ecomodernists are wrong to dismiss solar — today's solar panels could power North America with less than a quarter of a percent of the land, or about two thousand square feet per person (far less than we use for roads). A full 40 percent of our energy could come from rooftops alone.[30] And wind turbines can coexist with agriculture, already a massive part of our industrial landscape. Offshore wind displaces nothing of nature but (perhaps) someone's view, and in deep water, not even that. The ecomodernists' casual dismissal of renewables doesn't square with a position otherwise deeply committed to progress and innovation. Unless it's a nuclear advocacy group in disguise.

Break Through is a great book with a bold thesis. I agree with much of the "Ecomodernist Manifesto," especially the use of technology to contain our economic engine's bounds and the deep commitment to nature's independent value. But the leading proponents — Shellenberger, in particular — have morphed into full-throated defenders of nuclear[xiv] and natural gas who actively disparage anything else. The irony is best captured in *Break Through*: "Special interests are born when people reduce complex reality to single essences and then claim to represent those essences politically."[31] The effect of their bias is to undermine their core argument and discredit an otherwise nuanced take on the emerging relationship between humans, our economy, and nature.

There are lots of reasons economies slow down. And not all growth is good growth. We might not want more open-pit mines or slash-and-burn agriculture because we want to keep our forests intact, for example. And climate mitigation demands we end coal production. But none of that means we need to target economic growth itself. We just need clean growth. And where might that come from? Technology.

Techno-Optimists

Sir Francis Bacon long ago declared that opposable thumbs and a rational brain gave humans mastery over nature. Applied science, for Bacon, was the driving force in reshaping the world to our own ends. A modern variant on Bacon's theme is what I call techno-optimism, the idea that the natural propensity for humans to invent new stuff is so strong, and our creative abilities so open-ended, that we'll invent our way out of whatever pickle we find ourselves in, including climate disruption. From new energy systems to adapting to a changing climate, all we need to do is fire up human ingenuity and let it run.

---

xiv   My own views on nuclear and gas are developed in Part Two.

The beating heart of techno-optimism is Silicon Valley. One of my favorite places in the world is the ferry terminal in San Francisco. Ferries bring fresh-faced technology workers into a city that has for some time set the pace in an unstoppable and utter transformation in the way we work and live. The aggressive optimism that underpins the data revolution fills this place: the conversations, the venture capitalists, the stream of young, smart people intent on disrupting yet another industry with networks, algorithms, big data, and smartphones.

Toronto is fast catching up with San Fran. My home base of the MaRS Discovery District in downtown Toronto is also filled to the brim with the entrepreneurs, scientists, and investors who seek to rebuild our energy, medical, and digital media systems. Geoffrey Hinton, one of the founding fathers of deep artificial intelligence predicated on neural networks and formerly at Google, is chief scientific officer of the Vector Institute at MaRS. Walking around the MaRS complex, you'll hear all kinds of crazy conversations about new applications for stem cells, devices that interact with your brain in real time, deep data dives into genomics, and, of course, clean energy innovation. It's where the inventor of Hydrostor first came to me with the crazy idea of putting giant balloons underwater to store energy, and where I was first shown a way to make biofuel cheaper than gasoline.

I'm the last person to deny the powerful force of human innovation. It's what drives investments at ArcTern. But deploying clean energy innovation at sufficient scale to reduce climate risk will take more than a healthy startup environment, where venture capital and a bit of luck can make a winner. Cleantech ventures are nothing like web startups. Entrepreneurs can't disrupt the energy industry with a clever new app. Think Bombardier or Tesla, not Uber. It takes a lot of time and capital to build a company capable of competing in global energy markets. And the winners need to sell utility-scale equipment that can beat fossil fuels at their own game. The technology bit is only the first step in a long and expensive process.

Clean energy projects, like traditional infrastructure, are capital intensive and are built using debt finance. Whether it's a solar farm in Africa, an energy storage system in California, a biofuel factory in China, or even Klein's community-owned energy systems in First Nations territory, developers need to borrow money to build energy systems of any size. Even many of the growing number of homeowners who put solar panels on their roofs use debt financing. It's bankers who provide the debt financing that scales energy technology. Their backing, along with utilities and industry as customers, is critical to the success of new energy technology. That's a problem for innovators. Bankers, like utilities, are allergic to technical risk. They love old stuff, not new.

And that's largely why the best and brightest Silicon Valley venture investors tripped and fell when cleantech first became a fashionable sector in the early 2000s. They treated cleantech as if it were the same as any other technology. Like IT — hire really smart people and drive your way through market and technical barriers with brainpower and money. But it's not like IT at all — and they lost their shirts. Energy systems are unforgiving. You can't pivot to a new strategy or product if you've chosen the wrong thermochemical pathway. Or hire more really smart people to push your way past the second law of thermodynamics. It's a lot harder to bring cleantech to market than those venture funds were prepared for. And those Silicon Valley venture capital (VC) firms insisted on reinventing everything from the ground up instead of listening to what their customers wanted. Those problems can be solved with a better investment strategy (see "ArcTern's View" in Chapter Five).

Elon Musk, of course, is the exception that proves the rule. He has single-handedly taught Detroit how to build a modern car with Tesla, the utilities how to build a modern distributed utility with SolarCity, and NASA how to launch a rocket with SpaceX. And he gets climate disruption. A few more Elon Musks would go a long way to disrupt energy supplies and flows. New technologies and business models coming out of Silicon Valley provide a strong underpinning to Climate Capitalism. But it won't be enough.

Just south of San Francisco, Silicon Valley's techno-optimism runs straight into some hard truths. Central Valley is the most productive modern agricultural land in North America. It's a triumph of Bacon's mastery over nature. Its high-tech, irrigated, supercharged fields provide most of North America's fruits, nuts, and vegetables. A couple of years ago, it started to revert back to desert as the longest drought in memory rolled on. Eight hundred thousand acres went unplanted in 2015. As farmers desperately sucked up groundwater, the land collapsed, wrecking bridges and viaducts. As I write now, California is being deluged with rain the likes of which it hasn't seen in decades. The devastating fires of 2018 only made matters worse. As the weather seesaws between ever-expanding extreme limits, that farmland will become ever more unstable.

The question is not whether we can invent cool stuff that beats fossil fuels. We can and we are. The main challenge for techno-optimists is two-fold. First, you can't replace the Central Valley (or the wheat fields of Saskatchewan or the pastures of Texas) with hipster hydroponic urban greenhouses. Large-scale agriculture won't be disrupted with "ag apps." You can't water a desert with desalination plants. Or protect a coastline with microchips. Deep resilience in our infrastructure is more like old-school public investment and a lot less like shiny new tech baubles.

Second, the pace of change needed to head off catastrophic climate risk is not compatible with the long-term, lumbering nature of energy systems. Whatever compelling new energy technologies we might invent, we still need policies to scale them up much faster than the market can do on its own. That implies we need policy accelerants as much as technology itself. Just ask Vaclav Smil.

Fossil Fuel Fatalists

Vaclav Smil is a notoriously pessimistic man. A longtime thinker about energy systems, he rates our odds of success ditching fossil fuels in time to avert climate catastrophe as low indeed. His deep pessimism plays counterpoint to the techno-optimist. His view matches that of countless oil executives I've watched point to graphs that show energy demand climbing for decades. A shrug of their shoulders says, "What could I possibly do about this?" A fossil fuel fatalist is someone who believes fossil fuels are so deeply embedded into our lives that it's literally impossible to shove them aside.

There are three related reasons for fossil fuel fatalism. For Smil, history tells us it takes a very long time to replace something as big and central as energy infrastructure, too long for us to significantly reduce climate risk. For fossil fuel executives, there is no way policy-makers can hope to stop people burning what they've got to sell. For yet others, the economic dislocation of even trying to get rid of fossil fuels is too much to bear. In all cases, for better or worse, we appear wedded to them. However hard we might try, we are fated to ride off the climate cliff together. I think we need not go so meekly into that dark night.

Smil bases his argument on the historical record. His is an intellectually honest view: it took nearly a century for previous energy transitions, even when the new source is better, faster, cheaper. The shift from coal to oil, for example, or oil to natural gas. Our current efforts to shift from those to renewables will be no different. We're historically determined to take just as long. He's certainly correct that's what history shows us. And he's also correct to note — as many techno-optimists in Silicon Valley discovered to their detriment — energy systems change agonizingly slowly compared to other industries.

Others come to a similar conclusion based mainly on a rosy view of their self-interest. ExxonMobil notoriously argued that, while climate disruption was real and perhaps even dangerous, they saw no reason to be worried that any of their assets would be stranded. In other words, no matter that the dangers were real, policy makers would never be able to enact a stringent enough framework to stop them from digging up and selling *all* of their known reserves. Nor would new technologies, like cheap solar and electric cars, prevent them from doing so.

Both arguments are self-fulfilling. For Smil's argument to hold, he must assume the conditions under which the current energy transition happens are the same as those in history. If they're the same, we will fail. As for Exxon, their absence in the cleantech space,[xv] as well as their past immoral behavior in blocking climate action and awareness, are transparent efforts to make their wishes come true. The counterargument in

---

xv   See Chapter Four. The amount of investment the energy giants have made in alternative sources of energy is a pittance compared to both what they spend seeking more unburnable reserves (see section "Carbon Bubbles, or, the Great Fossil Fuel Stock Buy-Back") and what the venture community risks on cleantech breakthroughs.

each case is the same: a combination of policy and technological development can ensure we are neither fated to repeat the past nor bound to enact the self-interested dream of oil executives. This is precisely what Climate Capitalism does — change the rules so we don't repeat the past nor bind ourselves to the wishes of oil executives who refuse to read the writing on the wall.

A more subtle argument was put forward by Francis Fukuyama, author of *The End of History*: "If some of the more dire predictions about global warming are correct, it may already be too late to make the sorts of adjustments in hydrocarbon use that will prevent massive climate disruption, or else the adjustment process will itself be so disruptive that it will kill the economic goose that is laying our technological golden eggs."[32] Here, Fukuyama shifts from the classic fatalist position to a warning against trying. By forcing a faster change than would normally occur, we might damage our economic engine. While this argument has merit, it dismisses two real possibilities: not acting brings much more downside risk (d'oh!), and while there may be casualties, economies are always better off having gone through a technological revolution.[xvi]

In my heart of hearts, I've always been sympathetic to the fatalist position. Every time I see a powerful diesel Caterpillar at work on a construction site, or feel the powerful roar of a jet, I have a strong intuitive sense of just how hard it will be to replace this stuff. A trip to China a decade ago depressed me for months. I was in the industrial heartland of Nanjing and didn't see the sky the whole time, due to industrial smog. Getting to Nanjing, I passed dozens of cities equal in size to Canada's largest. I got a deep sense that China had only just turned on their industrial engine, had only recently hit the gas, and was just so damn big! The sense of momentum was overwhelming. How on earth could we stop this machine? Efforts felt positively Sisyphean.

But what's the point of that thought? How does it help? It doesn't.

---

xvi   Whether it's the internet, genetics, internal combustion engines, or factory automation, economies have always grown as a result of technological transitions — as painful as they may be to some individuals, firms, or sectors.

The fatalists have some logic and a great deal of intuitive despair on their side, but on ours, we have realistic hope. Humans made progress based on less than that in the past. Yes, replacing fossil fuels will be difficult, even temporarily painful. Therefore, the degree of market intervention must be historically unprecedented. That's the basic lesson. So what? We knew that anyway.

Fossil Fuel Apologists

No one in their right mind would deny fossil fuels are useful. It's obvious: they're compact, portable, fairly ubiquitous, and relatively cheap. They drive the modern economy and will be hard to replace. These statements are trivially true and are the starting point of any reasonable discussion about the difficulty in transitioning to a low-carbon economy. That's what underpins the fossil fuel fatalist's position. It's a false dichotomy to say we must see fossil fuels as either good or bad. They're both. That's why climate disruption is a wicked problem. But fossil fuel apologists demand you pick one side or the other. No nuance for them.

An apologist extrapolates from the obvious fact that fossil fuels are useful to justify an open-ended defense of burning more. This differs in an important way from the fatalist's position that we'll *inevitably* burn more. The apologist argues we *should* burn more, and those who say otherwise stand in the way of progress. It's a bit like moving from "salt is good" to advocating we all start shoveling the stuff down our throat. This only works if more salt always brings more benefit, which is clearly not true. To think more coal is always good makes the same mistake. It relies on an overly linear and simplistic idea of how benefits accrue. Sure, there are lots of early benefits from burning coal. That doesn't imply those benefits keep growing.

The apologist's basic argument[xvii] is as follows: fossil fuels brought enormous economic and human benefits. Those benefits outweigh potential harm (pollution and climate risk). Hence, burning them is a

---

xvii There are variants, but this captures the apologists' fundamental position.

good thing. Because burning them is a good thing, we should burn more since that brings more good. Hence, advocating limits to fossil fuel use does harm. There are two glaring mistakes in this argument. First, it misses the possibility that we can create those benefits another way, such as by using clean energy.[xviii] And second, it understates climate risk.

Climate risk, properly understood, kills the argument because the benefits don't outweigh the risk. It's as simple as that. There's usually some kind of dodge at this point. Most apologists — Ken Green of the Fraser Institute, for example — cherry-pick which pieces of science they wish to accept and ditch the rest.[xix] Bjørn Lomborg, president of the thnk tank Copenhagen Consensus Center, openly abuses[xx] climate assessment models that underpin climate risk. Ken Epstein (author of *The Moral Case for Fossil Fuels*) is similarly selective about his data. The strategy is wrong but simple: minimize climate risk from fossil fuels and focus solely on the benefits. This is an inadequate and self-serving approach. Less charitably, these folks are lying through their teeth.

Lots of other misleading arguments extend from this shaky foundation. Let's take a look at a few of the more egregious examples from

---

xviii By definition, an apologist can't admit other options — that's what makes them an apologist!

xix In a series of debates I had with Ken Green on CBC's *The Exchange*, a tangled knot of inconsistencies, typical in apologist rhetoric, emerged. Initially, he wanted to accept only empirical evidence and no climate models — which makes his thermometer and slide rule a better gauge of climate risk than the substantive analytical firepower of NASA. When pressed to use IPCC data instead of his intuitions and thermometer, Ken cherry-picked the single most conservative prediction of the single most conservative model within the IPCC framework (1.5°C) to estimate total warming this century (we're at 1.2°C now). When I pointed out how far outside the consensus his estimate was, Ken openly conflated climate sensitivity (how much the earth warms given a doubling of greenhouse gases) with total warming (the product of climate sensitivity times total greenhouse gases). When I pointed out that mistake, Ken still insisted warming would be mild — even though achieving mild warming requires that we drastically limit greenhouse gases, something he denies we can or should do.

xx My extensive criticism of Lomborg can be found in *Waking the Frog: Solutions for Our Climate Change Paralysis*.

Epstein's book,[xxi] which serve to illuminate the general strategy of the fossil fuel apologist: whatever the problem, the answer is always burn more fossil fuels. There are three main lines of argument. First, because previous predictions of catastrophe did not come to pass, we should ignore the current rings of climate alarm — and burn more fossil fuels. Second, because increased carbon emissions closely track measures of human well-being — including a drop in climate-related deaths — we should burn more fossil fuels. Third, since solar and wind have not yet stepped up to the plate, we can assume they won't — and burn more fossil fuels. Each of these arguments is wrong. Let's take them one at a time.

Previous cries of alarm didn't pan out, so we can safely ignore current warnings. Epstein cites as evidence news reports[xxii] of claims of impending catastrophe by NASA scientist Jim Hansen and journalist Bill McKibben. A single extravagant quote is no more counterevidence of climate risk than a cool day in July. Professional risk management doesn't entail picking out a data point or quote you happen to like. Crop insurers don't estimate weather risk by saying, "The last time someone said there would be a drought, it rained!" Climate risk is gauged by the preponderance of peer-reviewed evidence gathered over decades by experts across many disciplines. Consensus expert opinion as reflected in groups like the IPCC[xxiii] underwrites climate risk — not Epstein's flippant selection of extravagant headlines.

Since carbon emissions are correlated with increased well-being, we should burn more of the stuff. There are two glaring errors here. First, carbon emissions are only indirect evidence of the harm fossil fuels do and are displaced significantly in time. It's not the emissions we're worried about, it's

---

xxi Given Epstein is a self-declared "practical philosopher" who practises "critical thinking," his book is a uniquely spectacular example of sloppy thinking.

xxii He also points to the old canard that back in the 1970s a few published papers predicted global cooling. So what? Lots of science papers have made mistakes over the years, like Lord Kelvin's famous calculation that the earth and sun were fifty times younger than we now believe. Science, unique among disciplines, progresses.

xxiii I'd point out the Intergovernmental Panel on Climate Change (IPCC) is the largest gathering of scientists working on a single problem in the history of mankind.

the altered climate those emissions bring about, which occurs significantly later than the emissions. Second, he confuses correlation for causation; it's not carbon emissions that bring well-being, it's the energy those emissions represent. Which is why Epstein also needs his final argument below.

Since solar and wind haven't yet stepped up to replace fossil fuels, we can safely assume they won't. The fact that solar and wind have not yet dominated energy production is irrelevant; all that means is we've only just started on the job. If I have a marathon to run, and someone spots me only one kilometer in, does that mean I'm not going the rest of the way? I'll speak to the emerging technology piece in Part Two. But for now, I'd point out that looking in the rearview mirror is not the best way to gauge where you're going.

All of Epstein's mistakes can be characterized as errors of inductive logic. The philosopher David Hume pointed out the problem of induction way back in the eighteenth century: just because every swan you've seen so far is white, does not mean all swans are white; a single counter-example upsets the hypothesis. Best to find independent reasons as to why swans are white to buttress the argument. In each of Epstein's arguments, he takes a tiny piece of data and uses it — incorrectly — to make broad claims. A few reports of failed predictions imply all predictions are invalid. A few examples of emissions correlated with well-being means they are causally and inextricably linked. A few years of solar and wind being outcompeted means they always will be. Very sloppy thinking for a self-proclaimed philosopher.

Furthermore, these are not independent arguments. Epstein needs all three to hold for his conclusion. If *any* of the following three statements is true, his house of cards collapses: climate risk is significant; the harm from carbon emissions is not instantaneous; clean energy has a real shot at powering the developing world. All three look true to me, although just one does the trick.

Let's take a closer look at another apologist classic. The Fraser Institute's Ken Green argues we need to let the developing world burn lots of coal because that's the best way to get people out of poverty. Since increased energy use is correlated with increased economic activity, and coal is an easy source of energy, then limiting coal keeps people poor.

This is riddled with mistakes. First, there are many antecedent conditions to poverty; people are poor for many reasons. Many inner-city Americans are poor, yet they live in the most energy-dense economy on the planet. Second, giant, centralized coal plants may not be the best way to get energy to those who need it most. Perhaps millions of solar panels are a better way to go. Third, there are lots of other ways to help the poor, if that's your main objective — increased education, microloans for businesses, laws that provide social justice, etc. There's nothing like an attack on coal to make the far right care about poor people.

Even when apologists are forced to accept climate risk, they still conclude that we should burn more fossil fuels. Why? Because coal makes us strong and better able to weather the coming storm. Adaptation is a better deal than mitigation. This is precisely the line Ken Green takes in our television debates when, under duress, he's asked to accept the hypothesis that climate risk is real. Under what scenario is it plausible to argue adaptation is a better deal than mitigation? Only when you have an accurate long-term cost-benefit model that puts adaptation in the front of the queue. Yet Green consistently argues climate models can't be trusted. You can't have it both ways — accept economic estimates that go your way and reject those that don't. If he cared to check, every credible report on the issue — the Royal Society, IPCC, the World Bank — agrees the cost of adapting to a hot world swamps the costs of mitigation. But that's not the conclusion Green wants, so it's not an argument he's going to hear.

That the arguments taken together are incoherent isn't important to apologists like Green or Epstein. What's important is to always land on the conclusion you want. Fossil fuel apologists are guilty of a constructed affect bias. Affect bias means we're more likely to believe an argument if we agree with the conclusion. By constructed affect bias, I mean apologists construct arguments with the express purpose of coming to a conclusion they like. A key test is whether any data or argument would convince them otherwise. I've never found that to be the case. Hence, either their thinking is faulty or they're lying.

It's true fossil fuels are useful, and they will be hard to replace. It's equally true Climate Capitalism must accommodate the energy needs of the developing world. But that's the beginning of the story, not the end.

The difficulty in replacing them sets a baseline for our ambition, not a limit. Wanting to burn more is like taking another drink in the hopes of delaying or avoiding a hangover. It doesn't work. Believe me, I've tried.

There's a more interesting variant on the apologists' argument. It's likely fossil fuels of some kind are a prerequisite for any technologically advanced civilization — defined as abundantly electrified — anywhere in the universe. The argument goes like this: absent sufficient easy-to-access, energy-dense material, no society gains a foothold on the technology ladder to advanced energy systems, because only lots of excess energy provides the free time and capacity for engineering development. Renewable energy, like solar PV, involves highly advanced technologies. Hence, you can't get to clean energy without first going through dirty energy. Fossil fuels are like training wheels. A test of an advanced civilization is whether or not they can take them off!

Technological Tribalism

The folks at the Breakthrough Institute love nuclear and natural gas and dismiss solar. Bill McKibben and the folks at 350.org do the opposite. Mark Jacobson of Stanford builds a case for 100 percent wind, water, and solar (WWS) in his Solutions Project, preferring that combination over large hydro and nuclear. He's taken a lot of heat for it, particularly from the nuclear industry. The "clean" coal folks think we can keep burning the stuff by capturing the carbon and storing it (CCS — carbon capture and storage), yet Vaclav Smil makes a good case it doesn't stand a chance.[xxiv] Radical thinkers like David Keith tout geoengineering and giant carbon-sucking machines as necessary Band-Aids to give us more time, while others think any discussion of planetary engineering is taboo.

It's like a classic scene in an old *West Wing* episode. The White House wants to hear about clean energy alternatives. The respective lobbyists

---

xxiv To store just one-quarter of the stationary sources (big stacks) of carbon dioxide in the United States, we'd have to build an infrastructure that carries and buries the equivalent of the entire world's oil supply. That's because the O_2 in CO_2 (roughly) triples the weight of an energy plant's output, compared to its input.

line up to make their case. The solar guy trashes wind. Wind trashes solar and geothermal. Geothermal denounces wind and solar. The scene ends with the White House aides looking exasperated, saying, "Who needs fossil fuel lobbyists with that lineup?"

It's true each technology has a downside. Solar needed lots of subsidies to get off the ground, and is intermittent, just like wind. The amount of WWS needed to power a country is massive and will take more space than nuclear. Nuclear is expensive, takes a long time to build, and needs to overcome public fear. Natural gas is not a zero-carbon source and has the added climate risk of leaked methane from fracking. Large-scale hydro means lots of flooded land and transmission lines. To operate at a meaningful scale, CCS needs massive and expensive new infrastructure

But let's look at the other side of the equation. Nuclear is a huge and muscular low-carbon energy source, and next-generation nukes are much safer and able to burn existing waste. WWS is theoretically capable of powering the planet and provides free energy once the plants are paid for. Natural gas is an easy and fast replacement for coal with half the carbon emissions. Effective CCS means we don't have to turn off thousands of existing coal plants. Large hydro has proven itself as the current clean energy champion by providing enormous amounts of reliable and low-cost power for decades. All these technologies have a role to play in mitigating climate disruption in some form.

How do the internecine battles between them take shape if they could all be part of a solution? Take Jacobson's WWS approach. His Solutions Project provides a breakdown of the wind turbines, solar panels, and hydro plants needed to replace fossil fuels in the U.S. and Canada, with details down to the state and province level. Its rough-and-ready analysis demonstrates the possibility of going fossil-free by 2050. The Canadian numbers[xxv] come out to just over half wind, a quarter solar, a bunch of hydro, and a tiny bit of wave and geothermal. His critics (at least those

---

xxv   In detail: 58 percent wind; 22 percent solar; 16 percent hydro; 2 percent wave; 2 percent geothermal.

who agree on the climate issue) argue it's naïve to think we can do it without nuclear.

Why does Jacobson dismiss nuclear? He believes it's not only dangerous, but twenty-five times more carbon intensive than wind. It's a ridiculous claim, based on an absurd chain of reasoning[33]: nuclear power leads to proliferation of weapons, weapons lead to nuclear war, and the carbon footprint of a nuclear war is massive. So, Jacobson's critics argue for nuclear to reduce the size and complexity of the renewable energy rollout while Jacobson avoids nuclear because he's constructed an argument that predetermines only renewables.

The truth is we need them all. No single technology pathway is sufficient. Nothing should be ruled out in advance. Want to go all nukes? Canada needs a hundred new plants. The U.S. needs at least a thousand. That's two new plants every week for fifteen years. It takes a decade just to get these things approved. Want only large-scale hydro? We haven't got enough sites left to develop. No matter which route we go, the roadblock is scale — it's a huge undertaking. No single technology choice will change that. This shouldn't be surprising, since fossil fuels have powered our economy for several generations. Replacing it all in a few decades isn't going to be easy.

The real competition for one clean energy source isn't another — it's fossil fuels. We can (mostly) all agree on the primary outcome we seek — lower emissions — and set market conditions to get there. Secondary outcomes can also be valued: grid stability, an emphasis on capital costs over operating costs, redundant supply, etc. Instead of bickering, let's have our favorite horses race in a well-defined market. Venture firms will make bets. Energy companies and utilities will respond. Individual technologies will win in some jurisdictions and fail in others. Some will notch early wins but diminish in importance as energy systems evolve.

Predetermining the outcome by backing a favorite and trashing the rest underestimates the ability of the market to sort out those complexities in situ, in real time. Technology tribalism — backing one form of low-carbon energy over another — is not helpful. Everybody's got a favorite, and it's not a big step from defending its viability to trashing the competition.

Carney: Banker Rebel or Environmental Prophet?

Mark Carney is no tree-hugger. But the current governor of the Bank of England and ex-governor of the Bank of Canada sure sounds like one sometimes. He speaks openly about the "catastrophic impact"[34] of climate disruption as "a defining issue for financial stability." His comments are what you'd expect when you cross a climate hawk with a central banker: a polite, measured tone warning of impending catastrophe and stranded assets.

Carney's job is to manage risks that threaten global financial stability, and make no mistake, despite Brexit, astronomical levels of Chinese debt, and Greek euro-troubles, he believes climate disruption to be the single greatest threat. By speaking to financial markets in a language they recognize and from a position they respect, he's working to normalize climate risk as a fiduciary concern of financial professionals and make coping with it everyone's issue, including all those quants in the City and on Wall Street and Bay Street (in Toronto). One hopes this will ultimately result in unlocking the trillions of dollars of capital we need to lower the heat.

Environmentalists traditionally speak of climate disruption as a "tragedy of the commons," the difficulty of aligning self-interest to the common good in a system with shared resources. Why pay to lower your own pollution, when everyone else shares the benefit but not the cost? We know how to break that impasse, as politically difficult as it might be: we impose laws that limit emissions either by pricing the externality of carbon or through other regulatory mechanisms (the subject of Chapter Three).

Carney sees the problem from a slightly different perspective. He speaks of the "tragedy of the horizon" — the problem sits over the time horizon of most financial and political actors. From the perspective of those best able to provide solutions, climate risk is unseen — the business cycle is year to year, if not quarter to quarter; the political cycle is a few years at best; even central bankers like Carney think maybe a decade out, at most. Once the effects of climate disruption are felt, and the storms come over the horizon, it's too late to do anything other than hunker down.

For Carney, the answer is to engineer signals that affect financial markets today, but are tuned to be sensitive to events over that longer time horizon. By linking financial stability to the various risks that climate disruption will bring, he makes climate action something central bankers have to worry about. And that gets the attention of the financial community. He's providing reputational cover for financial professionals to take action on something many understand is critical but few know how to talk about at work or act on alone.

What are the risks Carney sees? Physical risk, of course — the usual floods, fires, and droughts. But he adds two others that engage the financial community along new lines. Liability risk for fossil fuel producers (and their insurers) derives from the possibility that those who suffer physical damage or loss may sue for damages. But the biggest is what he calls "transition risks — the financial risks which could result from the process of adjustment toward a lower-carbon economy." For example, pension funds sitting on lots of fossil fuel investments could see their value evaporate as new technologies and policies come into play.

That there might have to be limits to production is something the majority of the fossil fuel sector simply cannot or will not admit to. The current value of fossil fuel companies is at complete odds with the stated goal of world governments to limit warming to 2°C. As Carney notes, "That budget amounts to between a fifth and a third of the world's proven reserves of oil, gas, and coal. If that estimate is even approximately correct it would render the vast majority of reserves 'stranded' — oil, gas, and coal that will be literally unburnable without expensive carbon capture technology, which itself alters fossil fuel economics." Until Carney spoke up, talk about the carbon bubble was the prerogative of NGOs and the odd ethical investment fund. With Carney, the risk of stranded assets went mainstream.

The implications are not just that some investors will lose money. What worries Carney is that, as the broader herd of investors suddenly gets hip to these risks, there will be a rush to the exits. A "wholesale reassessment of prospects . . . could potentially destabilize markets, spark a pro-cyclical crystallization of losses, and a persistent tightening of financial conditions." This reiterates a point I made earlier: we're at the point now where the action required to mitigate climate risk is getting severe enough to bring its own economic risk. An abrupt resolution to "tragedy of the horizon" is itself a financial risk. The cure is getting as bad as the disease.

Carney wants to engineer a soft landing, but he doesn't concern himself with policy. As a banker, he wants only informational transparency so markets have the *potential* to be rational, to be *able* to react to what's over the horizon: "A 'market' in the transition to a 2°C world can be built. It has the potential to pull forward adjustment — but only if information is available." Mandatory standardized reporting on climate and carbon exposure forces the investment community to place their bets on one side of the risk or another. Those who prefer to ignore it will have to put their money where their mouth is. Presumably, the wisdom of the crowd — and the fiduciary obligation to be careful with clients' capital — sensitizes capital markets to "over the horizon" risks.

At a dinner organized by *The Walrus* magazine at a posh restaurant in downtown Toronto, I sat next to the chairman of a large pension fund. My job was the usual one at these events: play the gadfly who provokes conversation with button-down Bay Street types. As affable small talk turned to the carbon bubble, the gentleman got visibly confused. Turns out he didn't know the difference between smog and carbon emissions. Cleaning a smokestack of smog does not cure climate disruption. That a person at the top of the executive team responsible for billions of dollars of long-term capital investments didn't know even the basics of carbon risk is inexcusable. Worse, he wasn't even curious. When I pointed out the difference, he became angry. "Why the hell should I care about that?" he muttered.

I love what Carney's doing. He's normalized a conversation long missing in lots of boardrooms. Pension funds have long time horizons. Longer than the typical corporate cycle. Certainly longer than the political cycle. Their fiduciary obligation to meet beneficiaries' needs goes out decades. They need to defend against exposure to high-carbon assets in a warming world. More importantly, I'd argue they have a fiduciary duty to invest proactively in solutions and advocate for the policies that get us there. Pension funds have some obligation to ensure a healthy economy, on which they rely. At a bare minimum, pension fund managers — and that includes grouchy chairmen — need to be aware of these risks.

Carney's proposal is just a way to force financial markets to abide by the old adage "a stitch in time saves nine." But I've my doubts that information alone is sufficient. Markets are moved as much by animal spirits — the emotions and instincts of us wonderfully erratic humans — as they are by anything else, even over reasonably long periods of time. And without clear, long-term policies in place, the risk will be understated — which Carney freely admits. What he's proposing is a necessary condition for Climate Capitalism, but by no means a sufficient one.

MAPPING IT OUT

From this plethora of competing views, we might map out what a pragmatic Climate Capitalism looks like. In true pragmatic fashion, it avoids throwing the baby out with the bathwater, and it takes what works from each of these arguments in order to arrive at a workable solution. Most of them have something of value to offer, and the weak Paris Agreement forces us to be realistic. Current levels of ambition, enshrined in that Paris accord and reflected in the short-term priorities of our business elites, are totally insufficient to keep us safe. The scope of the climate crisis demands we go beyond just trying to achieve the targets most of the world has already set.

Instead, we must somehow create positive feedbacks whereby early successes in reducing emissions translate to economic gain, which in turn leads to increased business confidence and political ambition to solve the climate problem. Climate Capitalism outlines a way to do just that — getting past current approaches that see emission reductions as "environmental compliance." Done right, we approach those efforts in a way that fosters a sense of competition for a growing and profitable clean energy market. That will ratchet up public expressions of ambition from self-interested businesses and poll-testing politicians alike. In essence, Climate Capitalism harnesses the competitive elements of the very market economy that got us into climate trouble in the first place. It creates a virtuous economic and political cycle with real potential to mitigate climate disruption.

For starters, let's acknowledge that Klein's broadside attack on neo-conservative market ideologues is spot-on. She's also correct to note that incremental change at this stage of the game isn't much better than climate denial itself. But also note that her preferred solution set — a smorgasbord of longtime leftist ideas — ignores the broad political consensus we need to generate. Political extremism on either side can't deliver radical economic change — the neocons will run us over the climate cliff in the name of preserving short-term profit, and the far left will have us arguing about wealth redistribution as a proxy to reducing emissions. The radical economic change required by Climate Capitalism does not need to be politically contentious; rather, it's best characterized as a centrist position, since it's designed to maintain as much of the status quo as possible.

The ecomodernists and techno-optimists can deliver on some of that consensus with their focus on technology and growth. Fossil fuel fatalists, while realists about the sheer scale of the problem, are wrong to conclude there's nothing we can do. We are only historically determined to continue fossil fuel use insofar as we refuse to fully exercise our will to act, to choose how history might unfold differently this time. Similarly, fossil fuel apologists are correct in pointing out the obvious — fossil fuels are useful — but disingenuous to the extent they willfully ignore alternatives and climate risk. From the technology tribalists, we get the need to focus on ends, not means. The market will best decide what technologies are best suited for what outcomes. Markets can provide nuanced signals — including putting a value on things like grid stability and reliability, in addition to carbon emissions — to more broadly define that suitability. There's no need to predetermine favorites, like nuclear, while denigrating solar (or vice versa).

Finally, there's Mark "Captain Climate" Carney. Radical economic change won't happen without buy-in from those who write the big checks — including grumpy self-interested chairmen who don't want to hear about it. Carney wants to make climate risk something no one can ignore. Clear information linking "over the horizon" climate risk to today's investment decisions means every quant in the City, Wall Street, and Bay Street has to take a view. Those who think the problem is overstated can put their money where their mouth is. Smart money will go long on cleantech and climate risk, and short fossil fuels. Dumb money will short cleantech and climate risk, while staying long on fossil fuels.

All of these views rest on prior assumptions or within some overriding theory. Klein's leftist ideology plays counterpoint to the apologists' and techno-optimists' overriding belief in free markets. Fatalists' historical determinism prevents them from seeing possibility. Carney thinks information alone will render the market rational. Ecomodernists fell in love with nuclear and can't seem to see much else. Each position seems more wedded to preexisting theoretical commitments than a shared goal, which serves to increase my suspicion of overarching narratives versus a collection of piecemeal, pragmatic efforts. However messy and theoretically unsatisfying they may be.

If we have learned anything from the climate debate, it's that the problem is not simple. Nor will the solution be. No one ideology has a monopoly on, or even a viable vision for, how to fix the climate problem. This isn't a black-and-white nuclear versus non-nuclear issue, nor one of free trade versus protectionism. Or grassroots activism versus the elites of Davos. We don't have time to wait for *la revolución*, nor for global billionaires to save us. There is no magic bullet or single policy choice. No political heroine. And certainly no ideal political or economic system that comes with all the answers. In the face of the biggest and most urgent issue humankind has ever faced, we need to act quickly to make the needed change. Climate Capitalism proposes working with what we have, and working together, to mitigate climate disruption before we run out of time. The questions become: *How* can modern capitalism solve our climate crisis? *In what form* might it be compatible with a stable climate?

CHAPTER TWO

CAPITALISM:
CAVEATS AND CRITICS

Capitalism means different things to different people. The word does yeoman's work as a stand-in for lots of related concepts, and it's packed with hidden assumptions. To the far left, it's a pretty good proxy for the embodiment of evil. Today's neocon sees it as a panacea for all that ails us, the single source of wealth and progress. It's confused with globalization, trade, deregulation, privatization, finance, democracy, neoliberalism, neoconservatism, market economies, laissez-faire markets, and even fuzzy notions of freedom.

In everyday language, it can mean something as simple as opening your own bakery, the ability to free yourself from the tyranny of employment and follow your entrepreneurial dreams. On the business pages of the *Wall Street Journal*, it implies unimpeded power of global corporate giants and unfettered capital flows across borders. In Silicon Valley, it's getting rich with options in a technology startup. To a working couple, it means a growing investment fund to provide a stable income in retirement.

The two biggest confusions are whether capitalism is the same everywhere (it's not) and whether it implies an absence of state activity in the market (it doesn't). There are as many flavors of capitalist economies as there are cultures and countries. It takes one form in Russia, another in

Singapore and Sweden, and yet another in the United States. Further, over the course of history, in any single country, it evolves in concert with the changing times, technologies, and customs. Monolithic notions miss this variance.

Nor does capitalism imply that all economic value is measured and traded in the marketplace. Volunteerism, choosing a more meaningful but less lucrative career, charitable giving, and even taking care of our kids and elderly parents are all forms of value not formally captured in monetary terms by traditional markets. But so what? That values exist independent of purely financial transactions is neither surprising nor concerning. What markets can and do capture is dependent on what society decides they must. Markets are responsive to values to the extent that the rules by which they operate reflect those values. And in a democracy, we the people make the rules.

Only over the last few decades since the collapse of communism has an extreme neoconservative, libertarian view of free markets, which eschews *all* state intervention, been thought to be some natural endpoint of economic and social evolution. With apologies to Ayn Rand acolytes, that particularly virulent version of capitalism has shown itself to be unstable, destructive, and logically incoherent.

Growth remains a central tenet of all variants of capitalism, as central to the Swedes as it is to Americans. The need for growth today implies a need for growth tomorrow, and hence endless growth. It is perhaps on those grounds that criticism of mainstream economics has the most bite. Of most immediate concern are the practical constraints already felt along a number of ecological limits — from growing water scarcity to the diminishment of productive topsoil — to say nothing of the more philosophical paradox posed by endless growth in a finite world. Climate Capitalism squares the circle by arguing, whatever the long-term theoretical conundrum, we have no choice in the short-term but to aggressively target green growth.

The quest for growth is not unique to capitalism, however. It appears part of the human condition. No country or economic system of which I'm aware, no matter how far left — from Venezuela to Ecuador to the Soviet Union — has ever eschewed growth. Anecdotally, nor do any of my less wealthy or more ethical friends. Almost no one turns down a raise, and the first thing they do when they get one is book a trip to a place they love. Capitalism is just more transparent about the central role growth plays in its worldview.

To those who have some predetermined view of the moral status of capitalism — as either good or bad — I have a proposition likely to be unpopular: there is no inherent moral or ethical dimension to capitalism. Just as human nature can be empathic or self-interested, inward-looking or outward, nice or nasty — so too, markets, profits, and money can exhibit almost any virtue or vice. Markets reflect the behavior of individuals but operate under rules imposed by the collective that reflect the concerns and values of citizens as a whole. Climate Capitalism eschews the idea of preexisting moral underpinnings in the general notion of capitalism. Instead, our analysis starts not from an entrenched position founded on what is right or wrong with this economic system, but from the problem itself — the climate problem — and looks to the landscape of potential solutions to pick what works to formulate a solution.

PURISTS BEWARE

When any of us speaks of capitalism, we bring baggage that reflects our own set of background beliefs, interests, and prejudices. Capitalist economies come in lots of flavors, but their shared essence is not complicated. It's just the way we've always done business together. Let's untangle this mess.

According to Wikipedia, capitalism is "an economic system based on private ownership of the means of production and the creation of goods and services for profit. Central characteristics of *capitalism* include private property, capital accumulation, wage labor and competitive markets." The Oxford definition is "an economic and political system in which a country's trade and industry are controlled by private owners for profit, rather than by the state." An ideology-free working definition might then be an economy premised on investing private capital in efficient markets to make a profit. Clearly, private ownership, market forces, and profit are critical. What does this imply for the relationship between the public and private sectors?

A purist on the right might argue for zero government interference to maximize market freedom. That's absurd. The state is required for *any* market to exist. It logically precedes the market. At minimum, we need laws to protect private property and contracts, and courts to enforce them. Markets do not arise spontaneously, but only with preconditions set by government. This has never been clearer than in the modern era, when global free trade agreements don't just say, "On your marks, get set — trade!" but contain thousands of pages of complex rules enforced through large, opaque, and powerful institutions like the WTO. Free markets are *creations* of the state, as are global trade agreements.

The very notion of "freedom" is fraught. One person's freedom ends where another's begins. My rights imply your obligations. These critical points are often ignored by those of a libertarian persuasion. My freedom to swing a fist ends where your face begins. My right to trial by jury obligates others to sit on that jury. These simple points are easily extended to pollution: I'm not free to dump garbage, because you have a right to be free of that garbage. Greenhouse gases are no different in principle from raw sewage. Freedom does not imply the right to do anything, and market freedoms are never absolute. Those limits, enforced by law and not mere common decency, are necessarily defined by the state.

Further, we restrain some kinds of activity to ensure markets are both efficient and stable. Antitrust laws prevent monopolies. Banks require laws to force them to keep adequate and secure capital on hand. A strong central bank rescues Wall Street from its own excesses. We've learned the hard way, more than once, that markets are not self-correcting. Without state oversight, they bring unacceptable levels of instability and economic risk.

Hence, there is no such thing as a functioning market independent of the state.[i] Deregulated markets are not natural things that pop into existence; they are legal and regulatory constructions of the state. Markets, without oversight, can be inefficient and wildly erratic. State-imposed structure is a logical requirement for markets to *exist*, and state activity a

---

i   The reader may want to visit Mogadishu or the badlands on the Pakistan-Afghanistan border to see what sort of markets exists absent an effective state.

precondition for them to be *effective*. It's just a matter of degree, and on that point, there are healthy disagreements.

As a practical matter, the state has always been a partner to private markets in creating the wealth we see around us today.[ii] The heroically independent wealth-creating entrepreneur is a myth. Certainly, entrepreneurs generate wealth when they bring technology to market. But they've never operated in a vacuum independent of state activity. Take that most iconic of technologies, the iPhone. Almost all of the really high-tech stuff it contains — GPS, touch screen, Siri, even the transistors themselves — was created with public funding. No publicly backed core research, and there would be no iPhone or soaring Apple stock price. And it's not just cool toys; biotech is underwritten by the massive public budgets of places like the National Institutes of Health, for example.

But the role of the state goes way past seeding technology with publicly backed research. The auto sector blossomed when the interstate highway system was built. The aerospace industry rocketed with military demand. The early internet was built by the military, with demand accelerated by the academic sector. Nuclear energy was the plowshare of World War II. One cannot imagine today's global fossil fuel industry without the ready availability of state armies to enforce or coerce supply. The state hasn't limited itself to research — the supply of inventions — but has played a critical role in creating the broader conditions that scale up demand.

To be clear, I'm not disputing that places like Bay Street in Toronto or California's Silicon Valley (or the oil fields of Calgary, or the financial centers of New York and London) create enormous wealth and innovation. But that picture is incomplete. It's like pointing out the Golden Gate Bridge operates perfectly well without all the scaffolding it took to build it. The scaffolding may not be visible on the operating bridge, but without it, the bridge couldn't exist. The state provides the economic equivalent of scaffolding on which private markets build what is only later seen as independent economic activity.

---

ii   See M. Mazzucato's *The Entrepreneurial State*.

There's good evidence the state has always been an integral partner in enabling private markets and supporting wealth creation. The degree and type of state intervention varies — early risk-taking, building supporting infrastructure, setting market conditions — and vigorous debate on the most effective forms of market participation is a good thing. But the purist idea that marks totally unfettered markets as a sacrosanct capitalist principle that single-handedly created today's wealth makes no sense — logically or pragmatically.

On the other side, purists on the left portray capitalism as inherently evil, unfair, or destructive. This is no less absurd, and reflects a shallow, one-dimensional view of markets and money. Most of the left's economic sacred cows have long been killed off. The pursuit of profit is not by itself evil, nor is private property. Few outside the extreme left believe centralized and detailed state management can match markets for their innovation, complexity, and efficiency. Public ownership of essential infrastructure — rail, water, highways — remains credible (even desirable), but the idea the state should retain ownership over industrial capacity is discredited. A central challenge to purists on the left is the evidence of massive uplifts in living standards across the globe that occur largely in well-run market economies, compared to the economic shambles that invariably result when governments overreach. Modern Venezuela comes to mind.

Acknowledging capitalism has flaws doesn't imply you have to be anticapitalist. Seeing libertarian free markets as incoherent and unstable doesn't imply it, either. Rather, both imply the need to correct for those flaws. There are no predefined limits on what those corrections might look like. Pretty much everyone agrees selling baggies of white powder to kids is a no-no. Is that a capitalist state? What about putting a price on carbon? Sales tax? Capital requirements for banks? Rules to prevent real estate agents from flipping properties? The same goes for public ownership of assets. If a local municipality owns the transit system, is it still a capitalist economy? What about sewers? Or utilities? Still capitalist? In Sweden, advanced education is free. Still capitalist?

All economies have rules that govern how business operates. Those rules vary from state to state and time to time. Ditching capitalism instead of changing the rules is like refusing to play because you don't like the game. Change the game. We're not bound by prior commitments to economic theory; rather, we're limited only by our imagination and the pragmatic intersection of public and private interests, of wealth and health. Those who urge we ditch capitalism itself — and, therefore, its market forces, private ownership, and the profit motive — have an obligation to explain what on Earth might replace it. I've yet to hear a coherent alternative. And I don't see any useful or politically possible line over which our economy will cease to be capitalist.

The left's criticism has real bite, however, as it applies to growing problems like power distribution, injustice, inequality, poverty, and environmental degradation. Sometimes portrayed as problems inherent to capitalism, the real target of these criticisms is the ideologically pure version pushed by neoconservatives of libertarian bent in which market freedom trumps other considerations. As we've seen, that view is incoherent and hence the target a straw man. The neocon notion that markets and morality coincide in some perfect harmony is as discredited a proposition as the idea we can operate the economy without market forces. Same goes for their disregard for a strong public sector.

There has always been a tension between money and political accountability. People with money can and do corrupt political processes to their own benefit. The Koch brothers have done it. As do lots of libertarian-capitalist billionaires.[iii] Privately owned media clearly shapes U.S. public debate to the (self-perceived) benefit of the billionaire class and detriment of public safety. Rupert Murdoch has likely done more to blunt climate action than any other person on the planet. This is not a problem unique to democracy, nor to capitalism. Money corrupts politics in Russia and China as much as it does in Trump's America. The left typically pushes against the power of big money and the right defends, or even amplifies, it. There is no easy solution.

Reasonable debate about how to prioritize and solve practical

---

iii   See *Dark Money* by Jane Mayer for a disturbing account of how American politics and public debate have been thoroughly hijacked by libertarian billionaires, whose work has been aided and abetted by tax law.

problems isn't a fight on purist grounds. It's a push and pull across a number of conceptual axes: social, political, economic, and even psychological. By highlighting tensions between human values and unregulated markets and defending a strong public sector, the left works to balance the narrow interests of a powerful economic elite. Market interventions reflect larger concerns than short-term profit. We embed moral principles through laws (can't sell drugs to kids), promote the common good (provide public education), protect the vulnerable (build a social safety net), ensure financial stability (banks can't do as they please), and fill multiple market failures (from making polluters pay to supporting basic research). Good public broadcasting, from NPR to the BBC and CBC, counters the self-interest of those media-owning oligarchs. Indeed, I think democracy is impossible in a country dominated by privately backed news feeds.

The division of labor between left and right goes back a long way. Following the French Revolution in 1789, the French National Assembly consisted of two main camps, the Third Estate revolutionaries and the First Estate royalists. The revolutionaries sat on the left side of the chamber. The royalists sat on the right. Ever since then, "left" has been associated with an activist government working to help the poor and fight for social justice. Critics saw naïve utopian idealism. The term "right" came to reflect social stability, a preference for incremental change to the status quo, and an emphasis on personal responsibility. Critics saw an ill-disguised defense of the privileged. Ever since, the left-right battle has largely reflected the makeup of that assembly.

The classic battle of ideas between British economist John Maynard Keynes and Austrian Friedrich von Hayek toward the end of the World War II reflected the same divisions in a new context. Keynes emphasized the need for governments to intervene in the economy to smooth out instabilities and ensure full employment using stimulus spending. Hayek saw government interference as the first step to tyranny. (A similar view put Alan Greenspan's knickers in a twist, at least until his Ayn Randian world collapsed in 2008.) Keynes won that round, but Hayek staged a spectacular comeback with the emergence of the modern neo-conservative movement centered in the Chicago School and personified in Milton Friedman. Recent stimulus versus austerity debates — characterized by economist Paul Krugman on one side and Mario Draghi (president of the European Central Bank) on the other — is a modern variant of that original argument.

This interventionist debate is invariably about macroeconomic cyclical patterns within the economy as they relate to full employment. At stake are core ideas about the nature of the economy itself: Does it tend to self-correct? Is equilibrium stable? Debates about market intervention to create Climate Capitalism are different in two ways, and they skirt between these more theoretical, deeply entrenched positions.

First, climate risk is existential. Its threat is external to whatever economic theory or framework one cares to defend or attack. The risk it poses is not cyclical but systemic and permanent. We're not arguing about whether we can (or should) ride out economic downturns from climate catastrophe as the economy retunes and recovers. We're trying to prevent a breakdown of the natural systems that underpin all economic activity effectively forever. Arguments about the natural state of the economy and whether it delivers full employment are irrelevant. In that sense, Climate Capitalism is a meta issue, defined and defended independent of one's view of traditional interventionism.

Relatedly, Hayek and Keynes were concerned about defending or attacking deep, structural properties of economic systems. Questions about the necessity of intervention or stimulus spending spoke to the nature of the economy itself. The theoretical stakes could not have been higher, if in practice they pale to climate risk. In that sense, the interventions required for Climate Capitalism are a much lighter touch. They pose no threat to either side's economic theory. The question is merely: How do we cork emissions? (Or, in more technical terms, how do we limit or capture the negative externality of carbon?) There is very little at stake in terms of the original interventionist debate.

One can link Climate Capitalism to the original argument about full employment and cyclical downturns. Indeed, I believe the most effective economic stimulus one may imagine is an aggressive response to climate. But that's a separate argument that must stand or fall on its own, the outcome of which is irrelevant here. The old left-right debate as manifest in interventionist policy to maintain full employment has nothing to do with the interventionism required to mitigate climate risk.

The left and right have for much of history seesawed back and forth, epitomized by the Keynes-Hayek debate. It is an interminable argument based on mutually incompatible assumptions about human nature and the natural state of market economies. As human affairs go, this is not such a bad thing. The left sees humans as empathic, and altruism is a core value. To the right, we are self-interested, and liberty and self-reliance are core values. To the left, markets are destructive and unfair. To the right, they are self-correcting bearers of wealth and liberty.

To purists on the right and left, I have a mutually unsatisfactory proposition: both positions have merit. Human nature is both empathic and self-interested. Markets, the profit motive, and private capital are powerful and creative forces without which the modern economy cannot operate; yet, untamed by an assertive public sector that reflects the concerns and values of its citizenry, those same forces are volatile, destructive, and deeply unfair. Markets on their own are certain to cook the planet, yet I cannot imagine how we lower the heat without them. Neither side has sufficient resources to answer the climate challenge.

It seems natural to start from a preferred ideological stance and work toward a position on climate. Our ideas about what's good and bad inform our judgment on how to solve a problem. Most positions outlined in Chapter One do just that. Climate Capitalism, on the other hand, starts from the problem and looks to the landscape of potential solutions to pick off what works.

We don't have any time left to rehash the same old battles between left and right. Nor will we resolve them anytime soon. The good news is we don't need to. What we need to do is come together in practical, testable ways that reduce emissions. The deep philosophical differences do not apply when it comes to Climate Capitalism. Points of disagreement — on trade, regulation, and growth — are each just potential tools in the climate fight, each to be judged on its ability to provide effective change over the next critical decade.

Our attitude to growth is a good example. To the right, endless growth is the answer to all that ails us. A rising tide lifts all boats. With more wealth, we are better able to care for the environment, address injustices, and so on. To the left, unrestrained growth is what brought trouble in the first place. And since endless growth in a finite world is an impossibility (and no panacea if unequally shared), it's a structural problem that can only be ignored for so long. A hotter world changes the terms of that debate. We need to figure out — and fast — how to operate the existing economy without greenhouse gas emissions. That means the only race for growth that really matters is happening in real time: the race between clean energy technology and global average temperatures.

GROWTH: BULLS AND BEARS IN ROME

Ever since the 1970s when the Club of Rome warned of the need for strict limits to economic growth, the debate about the relationship of the environment to the economy focused on innovation versus scarcity. On one side, techno-optimists (the bulls) believe innovation can push through any limit — when one thing gets scarce, we'll replace it with something else. On the other, eco-pessimists (the bears) point to the unforgiving math of a growing economy and population in a finite world — you can't innovate around the laws of nature.

Until now, this has been largely a philosophical debate. The modern economy has roared forward ever since the start of the Industrial Revolution. Resource bulls and bears fight it out during cyclical variations in supply, demand, and methods of extraction. As oil prices go up and the bears gain ground, fracking gives the bulls more time. The Green Revolution pushed through food production limits and put the agriculture bulls in charge. It remains to be seen how long-stressed farmland, supercharged with fossil fuel inputs, can keep up the pace. There is no clear winner yet. Each time the bears growl, the bulls ask for more time.

In one sense, catastrophic climate disruption from unchecked carbon emissions is a modern variant of a recurring theme. As we saw earlier, today's techno-optimists, the Cleantech Bulls, believe human ingenuity will bring low-carbon innovations to market fast enough to limit climate risk. Modern eco-pessimists argue climate disruption, the Climate Bear, threatens economic growth itself, regardless of technological advances in clean energy. But climate disruption changes this old debate in three crucial ways.

First, we have to constrain carbon emissions *before* fossil fuels get scarce. We're not limited in how much fossil fuel we *can* burn, but how much we *should* burn. Normally, scarcity increases price, which motivates the innovation that pushes through limits. Think of innovation as the market's natural defense system triggered by scarcity's price signal. But scarcity here comes far too late. Recall the carbon bubble telling us the market will burn us right through the danger zone. Long before we run out of fossil fuels, or see a price spike, our Mad Max future is assured. The market's natural defense systems don't work for climate limits.

Second, that means the only innate market signals that might wake up the bulls are the economic effects of a changing climate. As weather gets nutty, the resulting economic shocks might be the trigger that gets us off fossil fuels — right? Well . . . no. What those shocks trigger directly is defensive action to protect assets from the storm. When Hurricane Harvey flooded Houston, flood defenses got money, not renewable energy. Innovation in reaction to climate events will focus on adaptation, not mitigation — incremental changes to try and absorb those economic blows. Those shocks might trigger a *political* reaction, such as we saw at COP21, but that's not enough anymore. The political will has to be translated into market signals that are effective in waking the bulls — that stimulate *both* innovation and deployment of cleantech at sufficient scale to mitigate climate disruption. And that's going to take deep public sector intervention in the market to align our economic system with our most pressing policy priority.

Third, if we miss the mark on climate, we don't get a second chance. Traditional scarcity cycles play out again and again. The bulls always get another chance. The price of oil goes way up, and industry survives to fight another day. New agricultural limits would be tough, but we could still buy time by rewiring food priorities from, say, meat to plants. But once the climate gets unstable enough to matter to everyone, it won't stop heating for centuries. If we wait until it hits us hard enough to hurt, it's simply too late; the climate's got too much thermal momentum. Wake up the Climate Bear, give him a small advantage, and the Cleantech Bulls are not injured but dead.

Limits to growth are no longer an arcane academic discussion about the possibility of infinite growth in a finite world, argued over by economists with a philosophical axe to grind and lots of time to watch it play out. We've all got a stake in this bulls vs. bears fight. It's happening in real time, for very high stakes, with different rules. The market's natural defences are down, and we've got one chance to get it right. If we take a close look at how this fight is starting to play out, we can find a way to bend the rules to give the bulls a fighting chance.

Cleantech Bulls

Cleantech innovation is nothing like breakthroughs in IT. It's big cables transporting megawatts of power, not little wires pinging megabits of data; giant pots of hot liquid and superheated steam driving turbines, not miniaturized hard drives delivering video feeds; huge fields of solar panels in the desert, not millions of iPhones in people's pockets. Some cleantech is relatively nimble, like solar panels on your roof. That's one reason why SolarCity looked a bit like an internet startup (for a while). But that's a small piece of the energy pie; less than a quarter of solar goes to households. Cleantech Bulls need to drive innovation in massive, capital-intensive projects. They're not startups inventing new apps in someone's garage. To accelerate Climate Capitalism, we must first understand why the clean energy revolution is nothing like the information revolution that preceded it.

As a whole, the Cleantech Bulls are showing strength. Renewable energy is big and growing, even in the face of collapsing oil and natural gas prices.[iv] Led by a shining solar sector, it's no longer an outlier. Global

---

iv   Most renewable energy does not compete directly with oil since it targets the electrical sector, where prices are set by coal and natural gas. But the collapse in oil and natural gas prices has brought investor attention to renewables, particularly in Canada, where investors have been distracted for years by the massive capital requirements of an expanding oil patch. Those investments now don't look so good. Renewables are becoming a long-term hedge.

investment was above $300 billion[35] for the fifth year in a row, with huge growth in markets like China, Africa, Latin America, and India. That's six times the investment levels of ten years ago — real money, even to Wall Street. Most of it went to large utility-scale projects, ranging from hundreds of megawatts to almost two gigawatts.[v] Record new production capacity was added in 2018, breaking the hundred-gigawatt barrier for the first time. Total renewable energy installations outstrip fossil fuel installations year after year now. Renewables are showing that most aggressive of growth curves — exponential.

Solar hit an inflection point in 2008 when the price of panels began dropping fast, going down a staggering four-fifths in just six years. This was due mainly to high-volume production in China, which in turn was triggered by a combination of policy-driven demand in Europe and the Chinese state's provision of lots of low-cost (essentially unrepayable) debt to manufacturers. We owe a lot of gratitude to countries like Germany and Spain, whose expensive policies (feed-in tariffs[vi]) drove that initial demand. That policy started a positive cycle: scale dropped prices; solar began to compete without subsidies; which feeds more demand; and so on.

But it's not just about low-cost panels. Large projects need debt finance from banks. What also drove down costs is the interest rate charged on that debt. A conservative financial sector sees today's panels — invented in government labs decades ago — as "bankable" commodities: well-understood equipment with no technology risk. They're now old, that's why bankers love them and offer their lowest rates.

Compared to fossil fuels, solar and wind can also come online *fast*, especially in the developing world. India is getting aggressive on solar. They recently brought online their largest ever solar installation, Kamuthi,

---

v   A gigawatt (GW) is approximately equivalent to a large coal plant and can power about a thousand average North American homes.

vi A feed-in tariff (FIT) is a guaranteed price to energy producers, normally high enough to ensure a project's profitability. The price stays the same over the long term, even if the technology's price begins to drop, since it's based on the cost of that technology at the time it was installed.

clocking in at a massive 675 megawatts. Kamuthi was completed in under a year, a time frame that's unthinkable for a traditional energy plant (never mind nuclear, which can take decades). The panels pour out of Chinese (and soon Indian) factories. India hires thousands of people to install them on an empty field and hook it up to the grid. It's that simple.

We're beginning to see a virtuous cycle: developed countries provide the R&D that brings innovation; the private sector commercializes it; initial demand builds up through targeted public policy; that drives scale in manufacturing, which moves to low-cost places like China; and banks provide long-term debt because by that time it's no longer thought of as innovation. Subsidies seed the industry until that technology competes on its own and the training wheels come off. This is precisely what we saw in the microchip industry starting in the 1960s.

Electric vehicles (EVs) are also starting to scale, especially in the largest auto market in the world, China. Many European states and cities are proposing to ban internal combustion engines entirely in a couple of decades, a target that may have seemed unreachable just a few years ago. It's hard to see exponential growth in the early stages, but the signs are there if you look. Overall car sales in China dropped in 2018, but EVs were up. Volvo was one of the first major auto companies to announce aggressive EV plans: fully half their fleet will be all electric by 2025, and all cars coming out of its factories will have some electric component in the drivetrain by 2019. Even mining equipment, ferries, and giant Caterpillar backhoes are going electric.

Things are not so bright on the cutting-edge technology side. Venture capital for startups developing the next generation of clean energy technology recently hit a measly $5.6 billion, nowhere near peak investment levels before the financial crash of 2008. Government and industry R&D stagnated at about $28 billion. Part of the reason is Silicon Valley's miserable results in the early 2000s as the biggest venture firms attempted to duplicate in cleantech what they achieved in IT. There are reasons for their failure, as we'll see in Part Two. With the benefit of hindsight, the next generation of cleantech venture investors can expect to do better.

There's another wrinkle, which may explain the mediocre R&D and venture numbers. If the solar industry of the '80s was a single-lane track, it's morphed into a fast-moving superhighway. That highway runs so fast on old technology that it's hard for cutting-edge stuff to get on the road. We'll find out more about access to that highway in Part Two. But for now, think of an on-ramp as "having scale" and "being bankable." Without it, a technology can't get up to speed fast enough, and, however promising, it will likely stall out before making the fast lane. Companies building shiny new innovations face a chicken-and-egg problem: to get to scale you need to be bankable, to be bankable you need to be operating at scale. We'll need policy to build some on-ramps, or else the market will lock out the new innovations we need to keep bringing prices down and performance up.

There are exceptions. My fund ArcTern looks for technologies that leverage existing materials, processes, and infrastructure rather than inventing exotic new materials or machines. This shortens the time it takes for banks to get comfortable and lowers the complexity of manufacturing. Toronto's Morgan Solar[vii] reinvented the way light hits a solar panel. This is set to drive down solar prices even more. They don't rely on exotic new materials or manufacturing methods. Their innovation is finding new ways of putting together existing, well-understood materials and inserting them into existing solar manufacturing lines. That gives them a real shot at hitting the solar highway.

Big old utilities are under attack from the Cleantech Bulls. Innovations in energy storage, solar, and energy efficiency bring intelligence to a more responsive grid. Companies like SolarCity systematize installation and aggregate assets, which brings debt financing and lower costs. Smarter electronics such as Sparq[viii] mean solar panels can help stabilize the grid even while they feed in intermittent power. CircuitMeter measures in real time every watt used in a building, bridging data analytics and sensors to deliver smarts behind the meter. Batteries sit behind the meter in factories, allowing them to choose when to take power from the grid. These kinds of innovations threaten the traditional utility's business model. Some, like Arizona's utility and GOP governor, fight back with

---

vii   We'll hear more about Hydrostor and other technologies in subsequent chapters.

viii   Micro-inverters connect individual panels to the grid.

lawsuits to delay the inevitable. Others — like NRG Energy and Duke Energy in the U.S. and E.On in Europe — are rethinking their role in tomorrow's grid. For many utilities, it's join the bulls or get run over.

The Cleantech Bulls' holy grail is energy storage. Whoever solves that problem will change the way we make and use energy forever. Without storage, there's a severe limit on how much renewable energy we can stuff onto the grid, no matter the cost. With storage, we can go all renewable, all the time — whether or not the sun is shining or the wind is blowing. That storage is coming on fast, and it looks set to explode. Early movers — Germany, Ontario, and California — created initial demand. The U.S. market alone is expected to grow ten-fold, to four gigawatts in just five years. Globally, it's expected to double six times by 2030, exhibiting the same nonlinear growth we saw with solar.

Mega-entrepreneur Elon Musk made things exciting with his all-electric Tesla, powered by lithium-ion batteries pouring out of the massive Gigafactory in Nevada. That factory has the largest footprint in the world and is designed to be fully powered by solar and batteries. It's not an exaggeration to say Tesla single-handedly taught a reluctant Detroit how to build a new kind of car. Eventually, electric cars will do more than drive us around without gas. They'll provide millions of portable energy storage nodes. Then there's Tesla's Powerwall, a battery big enough to power an American home and elegant enough to hang on the wall. Tesla doesn't use exotic new battery chemistry. They use roughly the same Panasonic cells you find in laptops. There are lots of more exotic batteries coming, but again, it's the tried and true hitting the market fastest. Batteries have their own superhighway, just like solar.

My own favorite energy storage technology has nothing to do with batteries. Hydrostor stores energy in the form of compressed air in caverns carved out four hundred meters below ground, connected to the surface by a water-filled tunnel. Use electricity to run a compressor, send the compressed air down that shaft to fill the cavern. The weight of the water keeps the air in the cavern at a constant pressure. Sounds a bit nuts, but they're ready to build hundreds of megawatts of projects. Why? They use existing machinery and engineering techniques, which keeps the bankers happy. Critical to Climate Capitalism is the notion that bankers and financiers matter far more than innovators and entrepreneurs. And on cost, Hydostor wins hands down against batteries. We'll find out more about their approach, and also why bankers play such a decisive role in Climate Capitalism, in Part Two.

Directly replacing the dense, useful liquid fuels we get from oil is the hardest target for the bulls, but even here, they're making progress. Next-generation cellulosic fuels — liquid fuel from inedible plant matter (no corn, no soybeans, no palm oil, no crops that impact the food system) — are finally showing promise after years of expensive missteps. INEOS and Beta Renewables both operate commercial plants. Canada's Woodland Biofuels is next up. Their large-scale demonstration plant in Sarnia, Ontario, makes ethanol from wood and agricultural waste cheaper than the gasoline it replaces. They do it by ripping the fiber apart into small molecules, like threads from an old shirt. Then they reattach those molecules into fuel, like sewing those threads back into a jacket. As far as I can tell, they're the lowest-cost fuel producer in North America — including the incumbents.

But these Fuel Bulls are particularly hard to get out of the gate. The first commercial plants are expensive, costing upwards of hundreds of millions of dollars. That's an awful lot of capital for venture funds to put at risk. But until you've got that first large plant operating, high-risk venture money is pretty much the only game in town.[ix] Venture funds in Silicon Valley blew their bank accounts on some very expensive and public blowups that killed investor appetite. No one hurt the biofuels sector in North America more than Vinod Khosla, whose venture fund Khosla Ventures aggressively backed a number of failures, including publicly traded KiOR.[x] When KiOR flamed out,

---

ix   There are some public funds, but private money from pension funds or private equity is pretty much absent for that essential first plant.

x   KiOR went from lab experiment to large plant too quickly. Cleantech was fashionable for a few years in the early 2000s. As a result, companies with lousy economics and massive technical and scale-up risk could get backing. KiOR's output, a mixed oil that needed further refining, was expensive (more than $5 per gallon) even if their plant had worked as planned.

it snuffed out most remaining investor interest. Woodland is being smart, going to places like China and Brazil, where long-term strategic interest remains intact.

We can learn a lot from these Fuel Bulls. There are few venture firms left who invest in big-box cleantech — the stuff that needs to be built at industrial scale before you know it really works. Most funds that did cleantech in the past first cut their teeth on IT, and they blew their brains out in the early 2000s treating cleantech as if it were the same. It takes way more capital to prove out the technology, so a couple of failures can wipe out an entire fund. And unlike software, if you're committed to a thermo-chemical pathway that won't deliver the goods (like KiOR), you can't just hire a bunch of smart people to bulldoze their way through the problem.

There were once hundreds of venture firms across North America that listed cleantech as a priority investment sector. I can count the active funds today on one hand. We'll need to find creative ways to get more private, high-risk venture money flowing to something other than new varieties of Twitter and Instagram. It's one thing for ArcTern to pick off a few opportunistic bets that leverage existing technologies in a new way, but without lots of high-risk investments to develop exotic new materials and machinery, we'll be tackling a twenty-first century problem with twentieth-century technologies.

Private venture money and large-scale project capital won't flow to help the Cleantech Bulls until we've built on-ramps to the energy highway and found ways to mitigate the doubly high risk of infrastructure-focused cleantech that needs inordinate amounts of capital to scale. It's one thing to invent something, like a new energy storage or solar technology; it's something quite different to have that technology deployed at a scale that compares with fossil fuel infrastructure. We had the relative luxury of building existing energy systems over several lifetimes. Climate Capitalism is about helping the bulls get to an equivalent scale in very short order!

Overall, though, the bulls look strong. The low-carbon sector is growing fast, bringing huge economic opportunity for those who get it right. It employs more people in Canada than Alberta's oil patch, and more in the U.S. than the American coal industry — by far. Long live the Cleantech Bulls! But what are they up against?

The Climate Bear

The Climate Bear is already growling. Warming accelerated sharply in 2015 as oceans released heat they'd been absorbing for years, shattering global heat records. Opening shots typical of the bear are taking their toll: severe droughts and devastating fires like those across California in 2018; dramatic changes in rainfall, like the flash floods that hit Calgary and Toronto in 2013 and the U.K. and much of the U.S. in 2015; the horrific typhoon in the Philippines and hurricane in Puerto Rico in 2017; a rapidly melting Arctic; and ironically, the crazy cold in winter 2019 brought by an unstable Arctic vortex. All that is nothing compared to what the Climate Bear has in store. Think of warming as energy: we're adding the equivalent of 400,000 Hiroshima atomic bombs per day to the atmosphere, 365 days a year. That's an awful lot of energy the Climate Bear can use to wreak havoc.

And it gets fed more and more every year, thanks to the fossil fuel industry. Ever more energy gets trapped in our finite atmosphere, driven by investments in fossil fuels that dwarf what we dedicate to the Cleantech Bulls. We spend $6 trillion on fossil fuels every year, including *direct* subsidies of $600 billion. ExxonMobil and BP's own projections — which optimistically assume "governments will continue to gradually adopt a wide variety of more stringent policies to help stem GHG emissions"[36] — have us shooting well past 4°C. And they keep feeding their own machine with significant investment. Fossil fuel giants spent an estimated $170 billion exploring for new reserves over the last five years — reserves they intend to burn, but science says they cannot if we are to avoid catastrophic climate change.

At this point, it's not enough for the Cleantech Bulls to beat fossil fuels on price or performance. Recall Vaclav Smil: it takes nearly a century for new, better energy sources to replace the old. As carbon levels shoot past 415[xi] parts per million (ppm), a more appropriate question is: can the Cleantech Bulls step up *fast enough* to wrestle down the Climate

---

xi   Atmospheric carbon levels are climbing so fast, I've had to edit this number three times since I started writing this book!

Bear? That depends on how hard we try to build an effective economic system — that is, Climate Capitalism — to support, encourage, and accelerate the development of low- or zero-carbon alternative energy sources. Climate Capitalism is a clear answer to the problem before us. Free markets are not showing themselves capable of dealing with the Climate Bear, no matter how motivated cleantech entrepreneurs and investors might be.

The large insurers have already warned they may be unable to provide what most people take for granted: the ability to find reasonably priced insurance to indemnify them against the devastating financial impact of catastrophic climate events. Ernst Rauch, Munich Re's chief climatologist, says, "If the risk from wildfires, flooding, storms, or hail is increasing then the only sustainable option we have is to adjust our risk prices accordingly. In the long run it might become a social issue. Affordability is so critical [because] some people on low and average incomes in some regions will no longer be able to buy insurance."[37] The insurer of last resort is always the public purse, which has already had to step in to backstop real estate in Florida and help recovery in Houston, Puerto Rico, and Fort McMurray. The insurance community is giving early signals that the Climate Bear has better odds.

The bulls have muscle, but market forces are not enough to win the fight; those bulls need steroids. UN climate chief Christiana Figueres estimates we need to *triple* clean energy spending to stop the bear anywhere near 2°C. The aggressive market making of Climate Capitalism starts with a price on carbon, then accelerates capital flows with government-backed green bonds, aggressively low mandatory emissions targets, elimination of regulatory barriers, and market pull from militaries and governments. Which brings us to the state as a critical partner in this new capitalist model. That's the subject of Part Two.

As an investment class, Cleantech Bulls face a rapidly growing global market whether or not we feed them the steroids they need to win the big fight. Obviously, as an investor, I'm bullish on cleantech. But as a human being, I understand the odds are with the Climate Bear. It's one thing to build a profitable clean energy sector; it's another to build one big enough and fast enough to fight the bear. We're clearly not doing that. What keeps me up at night is not the health of the bulls, but the strength of the bear.

Back to Rome

The Club of Rome was largely dismissed at the time by most mainstream economists, but the prospect of unrestrained growth remains an open question. This is not unique to capitalism. There has been little serious engagement on the question from traditional economists. Partly, I think, because of a bias to techno-optimism, but mainly because we simply *have no idea* how to run an economy that isn't growing. All economies — including that of Chávez's Venezuela, the old Soviet Union, and modern China — endeavor to grow. The desire for more and better and new appears part of being human.

But it's the fact that growth is both assumed and required — not just desired — that makes it our Achilles heel. It's not as simple as us wanting more. That's a cultural or psychological trait, and is potentially subject to rapid transformation — change the way we think, change how much we want. But stability in any capitalist model requires growth. Without it, everything comes crashing down: no growth, no profit, no profit and private capital withdraws; no capital means factories close and spending on productive capacity stalls; that puts people out of work; which further hurts growth. It's a particularly vicious cycle at the heart of our economic system. It's why economies get so unstable if growth merely falters. Keynes's interventionist stance was predicated on governments providing stimulus at just such a time.

Government debt adds another ring of cyclic instability. Large amounts of the stuff are considered sustainable only with assumed economic growth over the long-term. A country's absolute debt levels can get high relative to gross domestic product (GDP) — more than twice GDP in the case of Japan, nearly 90 percent for Great Britain and just over 100 percent for the U.S. As long as the economy grows faster in real terms than the rate of interest the government pays on its debt, then the size of debt as a percentage of GDP will drop — even if tax rates remain constant. If not, a country teeters slowly but inevitably toward insolvency.

The interest rate at which governments can borrow is, in turn, linked to expected growth, since it's evidence of economic health. Higher interest rates in the absence of growth completes the negative loop, as Greece and its unfortunate citizens discovered in recent years.

Absent growth, economies are highly unstable. There's no intuitive reason a mature economy can't flatline: everyone remains employed and consumes the same amount. But that's not how it seems to work. As far as I can tell, economics has not figured a way out of the growth trap. In the past, new markets, a growing population, and the deep carrying capacity of the earth provided easy ways to grow. Times are changing. Population has to level off at some point. We've run out of new lands to settle. Ecological constraints are real. Everything now hinges on productivity, squeezing more value out of a given amount of resources. That's where technology and innovation shine. Productivity is what has kept the Club of Rome's dogs at bay for so long.

There will always be economic subsectors that grow, of course, while others shrink. Capital seeks those that grow and avoids those that don't. Whatever else happens this century, clean energy is one of those sectors with growth potential. But that's not enough. It's the growth of the economy as a whole that gets us a return on our pension funds and savings accounts. It's what keeps public debt manageable. That's why economists are so fixated on a growing GDP.

Real growth can be confused with inflation. The money supply is infinite, if the planet's resources are not. One can simply keep printing money, as many governments have been doing in the past few years under the guise of "quantitative easing." But that's illusory growth. A bond might keep paying interest, or the stock market keep rising and paying dividends, but if the real goods we buy with that money go up the same amount, it means nothing. The net return to those holding the stocks and bonds is zero.

So, *can* the economy grow forever? Optimists point to a circular economy driven by clean energy. Ever larger amounts of value are added by producing ever more complex *patterns* of material and energy. The earth may be a closed system with respect to materials (the odd asteroid excepted), but it's open in terms of energy. We are not an isolated system. There's no reason in principle that eternal growth under those circumstances isn't possible, at least on human time frames. Material can be reused again and again[xii] to form things of increasing value. That requires only technology and energy. The potential energy from renewable sources is effectively infinite, and it keeps getting cheaper and more efficient. The IT revolution is all about making digital media, like entertainment, bring more real value with less material. The *pattern* matters more than the *material*. This picture isn't just a utopian fantasy of an ever-improving, ever-growing world; in fact, it has become a mainstream view. It's the core assumption that underpins every traditional economist's worldview, implicitly or explicitly.

But is that what's going to happen? I doubt it. My own view is infinite growth is impossible. Not because over the very long-term, the math doesn't work, but because the limits we encounter happen in the real world and in real time, unconnected to the idealized abstraction of economic models. The pragmatic necessities of a messy world trump ways in which we model it. Ecological constraints will bite faster and deeper than we might expect.

Take dematerialization: the amount of raw materials needed to produce a given amount of value drops as patterns replace material. It's true that, in theory, all material could be infinitely recycled and economic value derived from the various forms that material takes. It's also true higher proportions of economic value can come from entirely non-material sources. Hollywood movies derive huge revenues from relatively little material. And luxury items like a next-gen iPhone or Gucci bag need far less material for a given value. Given clean energy inputs, and fully circular material recycling, the argument has merit.

But if dematerialization is real, then surely the first commodity to go would be paper. The IT revolution put newspapers out of business, our phones have become our default for consuming written material in many cases, and paperless offices are the norm. The result? Global paper

---

xii ArcTern invested in a company called GreenMantra, for example, which can turn plastic that can't be recycled into specialty chemicals like industrial waxes. The same idea can be applied to all raw materials, from steel to cement.

production keeps hitting record levels, year after year — now at more than 400 million tons annually. Production may eventually level off, but as the saying goes, eventually we're all dead. I see little evidence of dematerialization on a scale and at a pace that shows economic activity is decoupling from environmental stress in even the simplest of cases. Maybe there will be exceptions — distributed ownership of autos, for example. If self-driving cars become the norm, individual ownership becomes a burden, not a blessing. Why pay to buy and insure a giant machine that sits idle most of the time? There's an awful lot of steel, plastic, rubber, and glass sitting redundant in driveways and parking lots.

Current growth comes mainly from the developing world. What people want there is basic stuff. Long before we get to internet-connected recycled crystal salt shakers and digitally enabled Gucci bags, they'll want more meat on their plates, nicer homes, new televisions on which they can watch all those non-material Hollywood movies, cars to drive around in (even if they're electric), and so on. We can't undo material progress in the developed world or expect those in less-developed countries to go without it.

The world is not a theoretical place where models play out to some long-term equilibrium. It's a material place where events transpire in real time, with demands by people who see their expectations as reasonable: "It's not possible to go back once you've been going forward. From a philosophical standpoint I, an air-conditioning junkie, can't tell someone they can't have it."[38] Economic growth over the next few crucial decades will come from people in the developing world buying material things, like air-conditioning.

And ecological collapse will likely happen much faster than we think. There may not be time for innovation alone to save us again. Imagine living on bank withdrawals of $100 a day, which is more than the interest you earn on your capital. You eat the principal and exhaust the bank account. The bank responds by giving you a line of credit, pushing up your limit so you can continue to spend. Since you need $100 to live on, that's what you keep taking out. Maybe you get that down to eighty by tightening your belt, but you don't get down to withdrawing zero. You just keep getting deeper in the hole. Eventually, the bank will close the account. When that happens, you still need $80 to live on, but you get zero, not fifty or even twenty. Nothing. And you keep getting nothing, no matter how often you ask. Without a sharp eye on your balance, you have the illusion that your lifestyle will continue as normal and that change will be gradual, if it comes at all. Until the bank brings down the hammer.

That may be what the market is doing: disguising the balance in our ecological account by temporarily pushing up limits — not just on the value of a stable climate, but on soil productivity, on fish in the oceans, etc. — even as we head to systemic collapse (I'll address this accounting deficit in Chapter Six). We're already pushing against or past a number of planetary boundaries, operating limits beyond which there is risk of irreversible change. They include genetic diversity, biochemical flows of nitrogen and phosphorus, ozone depletion, freshwater use, ocean acidification, and climate disruption. We're living on the principal from all of these natural "accounts." It's possible ecological systems won't wind down slowly but collapse catastrophically. Innovation in that case deludes us. Remember the cod fishery? The last few years' catch of Atlantic cod were enormous, because fishing boats got better at finding the few remaining pockets. Then cod stocks collapsed.

I don't know how this will play out. Neither does anybody else. Maybe the Club of Rome is right, and maybe they're wrong. Ingenuity might push the boundary out further and further. Fish farms fed on insects grown in vast underground protein factories come to mind. Whatever the outcome, one thing is certain: in the next few decades, we will face economic constrictions imposed by an increasingly unstable climate. And those constrictions will likely keep getting worse for longer than industrial civilization has existed if we do not act with real urgency today. The general challenge of infinite growth might be something we have to face eventually, but climate contraction is a real threat now. Not tomorrow, not theoretically — here and now.

Klein argues this is precisely the time to confront these deep contradictions within capitalism by voluntarily limiting growth. But she's wrong, at least as far as targeting growth; this is not that time. Mustering the political will to sufficiently turbocharge the clean energy sector is hard enough. Trying to simultaneously sell the highly contentious idea that we voluntarily give up growth will sink any environmentally minded ambitions linked to it. Politically, it's a nonstarter; people need jobs and vote to increase their economic prospects, regardless of their progressive credentials. The repercussions of no growth for pensions and government debt levels ensure zero support from the professional classes. Limiting growth kills the possibility of a broad coalition of business leaders taking action on climate, to say nothing of the reaction of the developing world.

We do know for a certainty, however, that high-carbon growth is an economic dead end. The resulting emissions will bring economic collapse.[xiii] To give the global economy a chance to grow, we need — at a minimum — to decouple economic activity from emissions. That implies massive growth of the low-carbon energy sector, which is something the right and left can agree on as a practical matter in the here and now.

Capitalism is not unique in its need for growth, although it is more explicit about that need. Whatever you think of the Club of Rome's pessimistic view, to get through the next half-century intact, we must replace existing energy systems with clean energy equivalents. And fast. We must *shape* capitalism to accomplish that Herculean task, not reject it. If we fail in *clean* growth, we'll never get a chance to answer the more interesting philosophical question of *infinite* growth.

The task to which we now turn is how to grow clean energy as fast as we possibly can. What sort of actions — public and private — are most effective? To best answer that question, we continue to ditch the old polarizing positions of right and left and focus instead on what capitalism might look like in a carbon-constrained world.

---

xiii   See Part Three: "Welcome to the Anthropocene."

PART TWO

CLIMATE CAPITALISM: CARBON, POLITICS, *and* SOLUTIONS

"Power concedes nothing without a demand. It never did and it never will."

— FREDERICK DOUGLASS, ABOLITIONIST

"It's a funny thing about capitalism: money you lose by slowing down is always more important than money you've already made."

— RICHARD POWERS IN *THE OVERSTORY*

CHAPTER THREE

PRICE AND THE NEED FOR SPEED

Way back in 1989, Margaret Thatcher addressed the UN General Assembly. She delivered a starkly worded message to world leaders gathered there, calling climate change the "single greatest threat to civilization." Known as the Iron Lady, Thatcher was no woolly headed environmentalist. She was a military hawk, champion of the free market, and kindred spirit to her transatlantic peer, Ronald Reagan. Scientifically literate and an economic pragmatist, her words echo as a lost opportunity.

Economics has long identified its preferred solution to "externalities"[i] like carbon pollution. Had we managed to act three decades ago, this chapter would be disarmingly short: price carbon. Return the money to taxpayers. Ratchet the price up gently. Watch the market work its magic. Everybody wins, right-wingers like Thatcher (and, presumably, current Canadian Conservative federal leader Andrew Scheer and his provincial counterparts, Doug Ford in Ontario, and Jason Kenney in Alberta) get a market-based solution with no net increase to the tax base; left-wingers

---

i   An externality is an unaccounted-for cost (or benefit) that hits a third party who didn't choose to take on that cost (or benefit). Throwing out garbage for free, for example, is an externality because someone else has to pay for the cleanup.

get a market tamed to serve the public good; and economic theorists (and the rest of us!) get maximal efficiency.

Despite a carbon tax being characterized as controversial in today's hyper-partisan political environment, there's not much room here for reasonable people to disagree. Pricing carbon is the *sine qua non* of climate policy — it's the single most efficient and comprehensive way to reduce emissions across all sectors of the economy. Acted on in Thatcher's time, or even in Stephen Harper's Canada, a gradualist approach and gentle landing was still possible; a slowly rising price on carbon would have given the industry and consumers decades to adjust their behavior.

But we no longer have the luxury of so agreeable a solution. Too much time has passed. Much more disruptive change lies ahead, no matter what our choice. Take your pick: it's either voluntary deep and rapid changes to the economy or inevitable ongoing economic disasters linked to ever-intensifying climate crisis. Pricing carbon may be the most efficient solution and has the backing of any economist worth their salt, but efficiency isn't the only game in town. Speed matters more now. Much more.

In the spirit of pragmatism and as part of the Climate Capitalism approach, we need to find ways to short circuit those parts of the economy that are slow to respond and coerce or cajole corporate leaders and investors to act faster than they might otherwise. We can think of short circuits as intervening in normal market operations in two ways: to *increase demand* for existing clean energy solutions (market pull) and to *accelerate invention and commercialization* of emerging solutions (market push). While the solutions I present here reflect my expertise, clean energy, equivalent opportunities to close gaps can be found across any number of emissions-relevant sectors, such as forest management,[ii]

---

ii   By putting a value on existing functioning forests, like the Amazon, we might limit the appetite for deforestation. And by placing value on carbon reduction, we might stimulate the planting of millions of trees on what is otherwise marginal land.

agricultural practices,[iii] and even social behavior.[iv]

The argument to increase market pull is a strong moral and weak contractual one, since it focuses on reducing emissions to meet ethical or treaty compliance rather than being driven purely by market forces (though there are long-term economic benefits[v]). All countries, with the possible exception of the United States,[vi] have voluntary emissions reduction targets to which they've committed. The voluntary nature of those commitments makes the contractual argument weak. However, a much stronger moral argument can tighten those obligations; countries that refuse to make good-faith efforts to meet targets are not much better than rogue states, given what's at stake in our collective failure. They can, and should, be treated as such through institutions like the WTO, which is in a position to punish them (see "Sovereignty and Trade Agreements: A New WTO Climate Consensus?"). We simply can't mitigate climate risk without working together, and the moral imperative of success provides compelling motivation to ensure compliance.

The argument for market-push policies, though, can be couched entirely in terms of economic self-interest. The global market for cleantech solutions is large and growing, expected to reach trillions

---

iii   No-till farming limits the amount of CO_2 released into the air, allowing farms to become more effective carbon sinks. Feeding cows a derivative of seaweed limits their methane production (mainly through burps).

iv   Linking personal choice to systemic change means more than voting for climate policy. When we each choose to fly less, buy a heat pump, or bike to work, we don't act in isolation. We provide for tiny nodes of action that link to broader macro-changes. Social change is always a mix of top-down guidance (laws) and bottom-up choices. Those links become more effective when we amplify feedbacks between levels of action: codify lessons from early-movers, broadcast what best practices they discover, extrapolate personal carbon budgets to national budgets, etc.

v   Lowering reliance on fossil fuels through efficiencies, for example, is pure economic upside.

vi   Trump's rejection of the Paris accord doesn't come into effect until after the next presidential election, so it's not certain that the U.S. will actually reject its (voluntary!) commitments.

within a decade.[vii] Accelerating solutions into that market means wealth and jobs. Canada can create far more economic upside by aggressively chasing that market than we can by doubling down on last century's energy resources. To get a big piece of it, we must bring better innovative solutions to market, at scale, faster than our competitors (like Germany, China, and South Korea). One does not have to be a climate hawk to want the economic upside that comes with billions in cleantech exports. Indeed, our economic self-interest and environmental aspirations dovetail in boosting exports that lower other countries' cost of compliance.

The challenge for a resource-dependent country like Canada is the presence of large and powerful domestic incumbents capable of diverting the resources and attention of the governments of the day. The oil and gas industry captured almost $5 billion in one fell swoop, for example, when Justin Trudeau's government bought the Trans Mountain Pipeline in 2018. That's more in a single deal, directed at an already-profitable and powerful industry, than the entire emerging cleantech industry was allocated for a decade.

Incumbents don't disrupt. They get disrupted. They're too busy making money. Global energy systems are changing rapidly, and those changes will come to our shores whatever we do. Solar and electric vehicles from China will do more to harm Alberta's heavy-oil-based economy than any environmental group ever could. Our economic self-interest is best served if we ensure at least some of those disruptions come from within our own economy. But Canada's incumbents continue to co-opt the national discussion and scarce public resources to support their short-term self-interest. Market-push policies are a hedge against letting those fossil fuel incumbents define Canada's long-term economic strategy.

The emerging competitiveness of clean energy (and its growing threat to our incumbents) may appear at odds with a plea for substantive and multi-pronged market interference to accelerate their adoption. As one economist commented to me, "If your new cleantech

---

vii   Bloomberg New Energy Finance.

stuff really is better than what the oil and coal guys have, we're done!" It's not that simple. As a cleantech investor, I'm bullish, but as a human being, I'm deeply pessimistic. The paradox of financial enthusiasm and moral horror is part of the psychological makeup of any clean energy entrepreneur.

I tell investors in ArcTern that cleantech competes head to head against fossil fuels on an even playing field. Solar, as one example, requires the least costly new energy infrastructure in many parts of the world. The cost of energy storage is plummeting. Energy efficiency plays have higher paybacks than one gets from riskier equity investments. All of that is true. And those market dynamics are unstoppable, whatever the Trumps or Doug Fords of this world try to do. That's why as an investor; I'm massively bullish on clean energy. My intention with ArcTern is to make my investors a lot of money[viii] building clean energy technology companies that further accelerate our shift to a low-carbon economy. We, and other early movers in this market, will do very well if we make the right choices (and have a bit of luck).

But as a human being, as a father with an abiding interest in the long-term health of our ecosystem, I'm also aware that — however successful clean energy might be — we need at least an order of magnitude greater deployment of this stuff to have even a coin toss's chance of limiting warming to non-catastrophic levels. I'm writing this book, and making the case for accelerated policy, as that concerned human being. As to my own self-interest as an investor, it's up to others to judge whether I'm just "selling my book" or putting my money where my mouth is.[ix]

---

viii   I emphasize here: if one does not make money in building and delivering solutions to a problem in a market economy, then there is no way that solution will permeate any portion of the economy outside where one imposes it. The presence of profit is what will drive other, less altruistic, investors into the space. Indeed, I argue that the most effective thing ArcTern could do is have enough success that sufficient competition arrives to put us out of business!

ix   I will say this: there are easier ways to make money than running a cleantech venture fund into which one has poured most of their net worth!

EFFICIENCY, SPEED, AND POLITICAL COST

Imagine your house is on fire. Aside from wishing you had paid better heed to early warnings from the smoke detector, what really counts once it catches fire is time. Speed matters. The longer the fire burns, the more damage it causes, the hotter and more out of control it gets. Which in turn makes it harder to put out, causing more damage, and so on. Sounds a bit like those nasty positive feedbacks in our climate system!

Now imagine a fire truck roars up and gives you a choice of fire hoses. The most efficient one — which delivers the highest proportion of water to the fire at the least expense — is the equivalent of a low-cost, high-tech Israeli drip irrigation system. All the water gets to the fire, but it's measured in drops per minute. The least efficient hose on offer is expensive and leaks all over the place, wasting water and money. But it nevertheless manages to deliver a massive flow of water to douse the flames, measured in thousands of gallons per minute. For the panicking homeowners, the efficiency of the fire hose is far less important than how fast it pumps water; they care about its *effectiveness* at putting out fires quickly. The smart choice is clear.

In a crisis, we are willing to trade some degree of efficiency for effectiveness. Increased cost for speed. Exactly what trade-off we'd settle for depends on lots of things, not least being the value of the house and ferocity of the flames. Delays cost in the long run because of the increased damage to your house. Hence, saving money on the hose is a false economy. That's where we're at with climate. There are two reasons why we may be willing to pay more for faster climate solutions.

First, and most obviously, we have to take into account the value of what we're trying to save (the house). Delays in climate action mean more emissions and an increasingly unstable climate. The International Energy Agency (IEA) and others have been hard at work trying to figure out the potential cost of the incoming climate storm. The problem is there are no economic models that can take into account the sort of stuff that happens past 2°C–3°C of warming. It's all (educated) guesswork.[x]

---

x   See Part Three, "Welcome to the Anthropocene" for some of the best guesses.

I've argued elsewhere that since we're risking everything, numbers are simply the wrong language[xi]; instead of fussing over dollars and cents, we may want to just draw a line in the sand and say, "Whatever it costs, we will de-carbon by 2030, because the stability of our civilization is at risk!" Our politicians and business communities are not there yet, but the military folks are pretty close. Simply put, it's dumb to save a few bucks if it means increasing the risk of hitting Mad Max levels of warming.

Second, there are more subtle costs to delayed action. Every time we avoid the nominally higher cost of installing zero-carbon energy sources, we build new fossil fuel infrastructure to meet our growing energy needs. That sometimes can look cheap in the short-term, but odds are good we'll need to turn a lot of it off before the end of its economic life. This kind of stuff — gas plants, pipelines, coal mines — is built on the premise it will operate (and generate returns) for decades. The problem is simple: what we've already got running takes us past 2°C. New fossil fuel stuff goes way past that. So we simply cannot let it run out its normal operating life. The government of Canada's recent decision to buy the Trans Mountain Pipeline — on the premise their ownership increases the likelihood of doubling its capacity — looks particularly silly on this view.

Shutting off a perfectly good coal or natural gas plant may seem politically impossible now given how ferociously their owners are likely to fight to keep it open, but in a decade or two, as the public gets more frightened of what's in store, it's not unlikely. According to the IEA:[xii]

---

xi   My previous book, *Waking the Frog*, speaks to the futility of using cost/benefit analysis to justify climate action (see Chapter Four). A preference for quantified analysis, backed by models that fit into a spreadsheet, over qualitative assessments can blind us to severe, existential risks best characterized by the latter.

xii The IEA are hardly climate alarmists. Assembled in response to the OPEC oil crisis, its main job is to advise governments about energy security. For most of its history, that meant how to keep oil, gas, and coal moving. Prior to coming out of the closet on climate about ten years ago, the agency typically spent its time tabulating how much fossil fuels reserves remained, keeping track of where they were, and giving comfort to those who worried about running out of fuel.

By 2017, we project that all permissible emissions in the 450 Scenario[xiii] would come from the infrastructure then existing, so that all new infrastructure from then until 2035 would need to be zero carbon, unless emitting infrastructure is retired before the end of its economic lifetime to make headroom for new investment. This would theoretically be possible at very high cost.[39]

We can put numbers on these kinds of costs. Way back in its 2009 World Energy Outlook, the IEA warned, "the world will have to spend an extra $500 billion to cut carbon emissions for each year it delays implementing a major assault on global warming." The agency put it another way in 2011, writing that "for every $1 of investment in cleaner technology that is avoided in the power sector before 2020, an additional $4.30 would need to be spent after 2020 to compensate for the increased emissions."[40] In other words, for every dollar we avoid spending now, it will cost us more than *four times* that much later. How is that a good deal?

As of the writing of this book, it's now 2019, and global emissions are higher than ever. Shutting operating plants before the end of their life will be expensive. And you can bet the investors who get hurt will expect the public purse to make them whole. This happened in Alberta when legislation effectively knee-capped coal-based electrical production there. But those owners have no excuse not to be able to see what's coming, and their pleas for compensation should fall on deaf ears.

Aside from *efficiency* (cost) and *effectiveness* (speed), we may also want to consider a solution's *political cost* and *robustness*.[xiv] In today's toxic political dialogue, climate action remains contentious. While discussion

---

xiii This was the IEA's most aggressive scenario deemed pragmatic enough to have a reasonable chance of success. It was predicated on limiting GHG concentrations to 450 ppm.

xiv Here I follow (broadly) the work of Mark Jaccard, who has long emphasized the varying degrees of political difficulty in getting one piece of climate policy in place over another. He has proposed a measure of the "political cost per ton" of carbon reductions in an attempt to quantify this idea.

of climate risk may have reached maximum toxicity at the federal level in the U.S. under Trump, Canada is not immune. Putting any climate policy into play generates some degree of antagonism from a portion of the electorate, and there is only so much any government can do before the public pushes back. Therefore, it's worth asking how much political effort a policy takes to implement relative to other options (political cost), and how hard it is to undo if targeted in an election (robustness).

Compare an economy-wide carbon tax to a shift in building codes to make office buildings more efficient. Since everybody sees the carbon tax (and generally dislikes taxes) that policy comes at a *high political cost*. It makes an easy target in the next election, since running to reduce taxes always makes friends. Hence, it's *not robust*. The opposite is true of the altered building code. Hardly anyone will notice the policy (aside from industry insiders), and the costs are directly imposed on a very small portion of the electorate. Hence, it has a *low political cost*. Few will fight the changes if they are applied equally across the industry; it's unlikely a political campaign will get fought over heat pumps and insulation. Hence, it's *very robust* compared to that unpopular carbon tax. In general, it may be the case that flexible regulations[xv] come at a lower political cost and are more robust than signature climate initiatives like carbon pricing. This is the drum Mark Jaccard has been beating for some time now.

We now have four ways to rank climate options: efficiency, effectiveness, political cost, and robustness. With these in mind, we can better focus on a single, pragmatic goal: how best to turbocharge the shift to a low-carbon economy. Central regulatory command or economy-wide carbon price? Market signals over flexible regulations? Legally binding

---

xv   A flexible regulation is one that defines the desired outcome without prescribing the means to achieve it. For example, net energy use per unit area of a building can be defined, and the market can be left to find the best way to hit the target. Or transportation fuels can be mandated to be of lower carbon intensity (a "clean fuel standard") while fuel providers figure out how to get there. The former is much more flexible than the latter as there are many more possible pathways to lower energy use in buildings than carbon content of fuels. More on the topic of flexible regulations later.

national targets or voluntary commitments? Carrots or sticks? To answer these questions without getting stuck in the same ideological commitments and interminable differences between the traditional left and right, we need a new language. Fortunately, there is one: evolutionary economics. It's a new way to understand the economy, one that uses modern mathematical tools to provide better insight into how it evolves or changes over time — as we'll see in the next section.

CARBON PRICING: WHAT'S OLD IS NEW AGAIN

Carbon pricing comes in two basic flavors. In each, one thing is known (set by the government) and one unknown (discovered by the market). A straightforward carbon price sets the cost to pollute, and the amount of emissions reduced by that price is discovered by the market as people and companies respond in various ways to that price. In cap-and-trade, it's reversed: policy places a cap on emissions (which reduces over time) and the market discovers the cost of hitting that target. It's impossible to know both the cost and amount of emission reductions simultaneously. There are a lot more details to these two concepts, but that's the basic idea.

Putting a price on carbon is not the slightest bit contentious (at least to economists), as we'll see. But what you do with the money certainly is! The money raised can be returned to taxpayers (directly or indirectly), which makes the policy revenue neutral. Or, those revenues can be used for other government activity (climate-related or not), which makes it revenue-generating. Obviously, a revenue-neutral carbon price is *more robust* (that is, tougher to undo) than one that is not, since reversing the policy would mean putting an end to the money flowing back out to taxpayers to make it revenue neutral. It also comes at a lower political cost because there is no net increase in taxation. It should neuter arguments about "bigger government" or "cash grab," since neither accusation is true. A revenue-generating policy, on the other hand, comes at a higher political cost (you have to convince voters of higher tax levels), and is less robust (undoing it is easier because it means lowering taxes, always a popular move). But it can be more effective if that money is used to further reduce emissions.

Take Ontario's cap-and-trade system, which came into effect in 2017 under Kathleen Wynne's Liberal government. Emissions were capped, by sector, based on historical norms. Individual polluters are allocated a cap, and they may pay to exceed it. The permits to pollute are bought at an auction, which determines the price. In each quarterly auction for the inaugural year, those permits sold out, raising nearly $3 billion. That money was to be used to further accelerate emissions reductions: subsidizing building retrofits, providing risk capital for industry to try new technologies, supporting electrification of transport, etc. Vilified as a tax grab in the next election, it was a gift to the opposing Conservatives wooing an electorate exhausted by high energy and tax bills. One of the first acts of incoming premier Doug Ford was to scrap cap-and-trade in its entirety. The upshot: while admirably effective (if it lasts!), an economy-wide, revenue-generating carbon price comes at a high political cost and is not robust. Undoing it was an easy (if irresponsible) promise to a tax-weary electorate.

Compare Ontario's experience to British Columbia, where all revenues of an economy-wide carbon tax offset other taxes. For every dollar raised on carbon emissions, there is a dollar reduced elsewhere in payroll or sales tax. There is no net new revenue to government. That makes BC's carbon price highly robust: opposition parties who campaign against it must raise the taxes that were reduced to offset carbon revenues. No one has even tried. There appears to be a ceiling on political appetite to keep raising that price, however, even if it's revenue neutral. As the years roll by, the carbon price generally has to level off as consumers begin to feel the bite at the pump.

What to do with the revenue may be contentious, but putting a price on carbon is not. At least not among economists, where it has always enjoyed broad support, and for good reason. Putting a price on externalities to reduce public harm is one of the least controversial assertions one can make in economic circles. Indeed, traditional thinking has long identified it as the most efficient way to deal with unwanted externalities. The reason for this is simple: all market actors, faced with a known cost to pollute, are motivated to find any and all ways of reducing pollution, making it cost less than that market price. The carbon price is a ceiling, and the market will hunt for anything that costs less.

Conservative economic thinkers, from Hayek to Friedman, explicitly endorsed this market mechanism to reduce pollutants. Former Federal Reserve chair Janet Yellen recently put it this way: "From the standpoint of an economist, the most efficient way to tackle climate change is to tax emissions — to create a disincentive to emit carbon dioxide. It's the right solution to a problem."[41] Outside the loony fringes of libertarian think tanks, nothing in substantive economic theory argues against pricing carbon.

A newer economic paradigm with more modern conceptual and mathematical tools at its disposal goes even further than traditional conservative thinking. Evolutionary economics agrees carbon pricing is efficient but identifies and explains pragmatic real-world limitations as to its effectiveness. By grounding those gaps in an empirically leaning theoretical framework, we can extend policy options beyond the basics of carbon pricing in a structured, defensible, and politically neutral way. In fact, Climate Capitalism calls for economic solutions that are in addition to, and go far beyond, just carbon pricing on its own. Because the climate problem is so urgent, we need more than one tool to fix it. Not only is our house in flames, it's a four-alarm fire!

From the Old . . .

Let's get our bearings and start with the old, traditional economics that's long underwritten the right's commitment to free-market ideology. Not only is there no contradiction between traditional economics and carbon pricing, but it emerges quite naturally as part and parcel of the underlying theory itself.

On the traditional view, markets are best left alone to find equilibrium. Technically, equilibrium identifies the place in abstract space found by economic models if they are left to run for some (generally near-infinite) period of time. Conceptually, it's a special place where everything is optimized, including human happiness. According to the right, it's what happens when the economy is left to its own devices, free of unwarranted government interference. The organizing force that gets us there is greed: people acting in their own self-interest. That force is a bit like gravity, but it acts on people (not mass) and is felt in dollars and cents (not weight). Adam Smith's "invisible hand" long ago captured the basic idea: "It is not from the benevolence of the butcher, brewer, or the baker that we expect our dinner, but from the regard to their own interest."[xvi] Let's call this view the Basic Picture.

Mind-numbing levels of detail[xvii] have been added to the Basic Picture. Decades of intellectual effort have been invested in defending it from various attacks,[xviii] often with an express intent to preserve Chicago School–style notions of free-market fundamentalism. If equilibrium really is a magic place of maximum efficiency and happiness, and we get there by unleashing the forces of unfettered greed, that makes for a pretty strong argument to leave business (and rich people) alone. Lots of pundits and free-market think tanks dedicated themselves to defending just such a view. As a result, anyone advocating for market interference of any kind faces strong headwinds. That's by design.

However, that characterization of the Basic Picture is shallow at best, and unfair at worst. Even those committed to free markets have the conceptual apparatus to support — even require — carbon pricing. It lurks in the simple words "find equilibrium." The trick to making this theory work for climate is to *shift the point of equilibrium*. We do that by ensuring greenhouse gas emission levels have some influence on where the system finds equilibrium. That's where a price on carbon comes in.

---

xvi   Adam Smith, *The Wealth of Nations* (London: W. Strahan and T. Cadell, 1776), Book One, Chapter Two, p. 15.

xvii   Which I've written about in *Waking the Frog*, see Chapter Three: "Complexity and the Myth of the Free Market."

xviii To take one simple example: because humans are clearly not infinitely intelligent, they cannot maximize their self-interest, since that would require near-infinite real-time comparisons between possible choices. Hence, economists reverted to using the term "satisficing" our self-interest, which means making a practical choice that is "good enough." While this response is clearly sensible, it's one of hundreds of Band-Aids put on the core economic model that assumes equilibrium as a primary operating principle. A partial list of assumptions that turned out false and required a Band-Aid: things in short supply can be replaced by something else; time doesn't matter; there is no psychological momentum in preferences; there are no transaction costs; etc.

An externality, an idea first developed by Arthur Pigou back in the Roaring Twenties, is something not accounted for in market pricing, like carbon pollution. It's free to pollute, so there's no incentive to stop, yet it causes economic damage. That damage is external to the system, or person, that pollutes. If something is free, it's effectively invisible to market forces, akin to having no weight and therefore not being subject to the force of gravity. To bring an externality into the market economy — to internalize it, in other words — you attach a price tag. It's like assigning mass to an object in a physical model: it no longer operates outside the system. Instead, it is now under its influence like everything else. Of course, unless it already has intrinsic market value — which, by definition, an externality does not — then the price tag can only be generated by the public sector, the ultimate arbiter of market rules.

"Equilibrium" has value in our analysis as the primary tool endorsed by right-wing neoconservatives — as long as they are capable of recognizing climate risk. The idea in this context is that we can shift the equilibrium to a state such that it reduces climate risk. That is, by putting a price on the externality of carbon pollution, over the long-term, people will pollute less; that's the new equilibrium — less pollution. While climate skeptics and fossil fuel lovers might abhor the idea of pricing carbon, there is absolutely nothing about classical or even neoconservative views of markets that prohibits the idea. Indeed, Milton Friedman, champion of the dominant free-market ideologues in the United States, explicitly endorsed the notion of pricing pollution. He even did it on *The Phil Donahue Show*:

> "Yes, there's a case for the government to do something [about pollution]. There's always a case for the government to do something about it. Because there's always a case for the government to some extent when what two people do affects a third party . . . But the question is what's the best way to do it? . . . The way to do it is to impose a tax on the cost of the pollutants emitted by a car and make an incentive for car manufacturers and for consumers to keep down the amount of pollution."

Just as chemical equilibrium is shifted by the addition of a new catalyst (speeding up the chemical reaction, for example), or the equilibrium water level in a lake system changes with increased annual rainfall, so too will the equilibrium level of carbon emissions change as a result of the increased cost associated with those emissions (getting lower on average as the price increases). There is nothing remotely controversial about that claim.

The only way to defend against carbon pricing on this view is to privilege one equilibrium over another. In other words, to privilege the economic equilibrium associated with planet-heating levels of atmospheric carbon released by unfettered markets over others in which carbon levels are limited. And why would anyone want to do that? Well, maybe the neoconservatives, who view market freedom as sacrosanct. Their entire ideology rests on the assumption of a single, "pure" equilibrium, the best of all possible worlds — even with evidence that we are headed toward a nasty world. Why? Where does that assumption come from? The answer goes to the heart of when and where the Basic Picture emerged.

Back in the 1800s, when the hard sciences were beginning to build mathematical models to describe their areas of interest — physics and chemistry, for example — economics wanted to get into the game. Unfortunately, the math of the time was so basic that it wasn't possible to describe and solve a system as complex as economic activity, which includes all sorts of difficult-to-model phenomena, like technological change, psychological preferences with recursive loops,[xix] etc. The only way to make the math work was to assume one point of equilibrium, and, yes, to assume that one point was the best of all possible worlds (it achieved Pareto[xx] optimality, named after the Italian economist Vilfredo

---

xix   What people expect affects the market's operation, which, in turn, can change what people expect. The simplest example is the self-fulfilling nature of economic downturns: if enough people (and corporations) feel pessimistic about their economic future, they stop spending money. This, in turn, can cause a recession, which further exacerbates their negative outlook. The types and degrees of psychological recursion within the economy are open-ended.

xx   Pareto optimality is an optimal allocation of resources in the sense that any reallocation to benefit someone will necessarily be at the detriment of someone else.

Pareto). It turns out that a single, privileged equilibrium, which markets find when unfettered, is just a hypothetical artifact of out-of-date math! It's baked into the whole structure of this theory for no reason other than, at the time, the equations couldn't be solved without it. It has no independent support.

Challenges to free-market fundamentalism, mainly but not exclusively from the left, can now be articulated as an argument, not against the Basic Picture itself, but against the hidden sub-argument that there's only one equilibrium worth pursuing — the "natural," single, pure (and completely theoretical) one that business and individuals will find on their own, free of government interference. Typical criticisms — a sense of fairness, the distorting power of monopolies and vested interests — can then be reinterpreted as a need for the public sector to impose rules on market behavior to shift the equilibrium in some way: a welfare program to shift to a place of more evenly distributed wealth; monopoly-busting to achieve a pre-distorted state; political contribution limits to avoid the distortions of an undemocratic public sector; etc. Each can be argued on its merits. Carbon pricing is no different. Use it to shift the equilibrium and, therefore, reduce emissions.

The answer from those defending the Basic Picture has always been — wait for it — equilibrium takes time. This answer is inadequate, for two reasons. First, as a practical issue, time matters. In the case of climate, our house is on fire. In other cases, like inequality, waiting around comes at a cost to those who are hurting. Second, from a theoretical perspective, the "wait and see" approach can't be falsified. If the problem doesn't resolve, one can just ask for more time. An assertion that cannot be falsified is an empty one and not evidence of anything. In other words, one can use the excuse "not enough time has passed" every time there's a suboptimal outcome or a challenge to the approach. There's no way to prove otherwise, except wait for a (possibly infinite) passage of time.

As much as this general criticism repudiates the idea of the single, theoretical equilibrium as nirvana, it's *not a rejection of markets nor of capitalism*. There are any number of variants on capitalist systems, including those that take the concerns of the left quite seriously — epitomized by Sweden or Denmark, for example — and impose external constraints on the Basic Picture to accomplish specific policy goals. An equilibrium (or limit) imposed on the market, from a perspective external to it (such as the government), is perfectly compatible with market mechanics and capitalist economies. One can import moral imperatives into the model as artificial forces to engineer a new equilibrium that achieves policy objectives.

The core of the Basic Picture has merit: shifting to a new equilibrium (over whatever time) indicates a direction. Pricing carbon (to bring the externality into the economic system) will move the economy toward lower emissions (the new equilibrium that achieves the policy goal). Rules to ensure it's in everyone's self-interest to reduce emissions puts Smith's "invisible hand" to work for our collective benefit (instead of unrestricted carbon's "invisible boot" kicking our collective rear). But beyond that, the traditional economics cupboard is largely bare in the context of mitigating climate disruption.

My own view is equilibrium itself has limited value as an analytical tool for anything beyond the mundane. Perhaps there is equilibrium in the price of sugar, or the stock of Tesla, at 2:30 p.m. on a Thursday afternoon, but it's a transient thing, ephemeral in the ongoing complexity of the human world. There's little evidence that equilibrium in human systems lasts over long time frames — the world is just too complex. Equilibrium is stasis, a kind of death. It's a test tube where nothing more is happening, a lifeless and uninteresting place. And it's of little help in our climate conundrum — it defines an obviously desirable end point, but has little to say in detail about how best to get there. And time, which has infinite supply in the Basic Picture, is entirely of the essence when your house is on fire! At best, the conventional economic model endorses carbon pricing as a way to shift to a lower-carbon state.

The fundamentals of traditional economic models came into being a couple of hundred years ago — long before we developed modern mathematical tools like dynamic systems theory and the mathematics of complexity in the late twentieth century. Modeling the economy using the Basic Picture is a bit like trying to build an iPhone out of a chipset from 1980. Complex modern problems, like climate, require complex modern economic models to help resolve them.

To the New . . .

Everything of interest that we study, every pattern that exhibits complex behavior, operates far from equilibrium. That's why that thing is able to exhibit interesting behavior: it represents some dynamic unfolding of complexity fed by an external source of energy that keeps it dancing. The pattern exists *because* it operates far from stasis. And its complexity increases as it interacts with its surroundings. From cells to weather systems, from social networks to neural networks, and certainly the global economy — all are best described as complex, evolving, dynamic systems, each underpinned by a common base of modern mathematical analysis. Equilibrium is both local and temporary, and of little value in understanding how the system evolves over time. It's like looking at a company's balance sheet, which captures just one moment with no context about how that company's health (sales, debt loads, etc.) is changing over time. It's like a photo in a world of movies. But economic theory is not immune to more complexity. The new economic game in town is known as evolutionary economics. Call this the Modern View, which improves on the Basic Picture with better tools to explain why carbon pricing is a singularly effective piece of Climate Capitalism. Not only that, it also validates additional potential solutions to the climate problem. Essentially, the Modern View gives us a bigger and better hose with which to douse our raging fire.

Newer mathematical tools — complexity theory, dynamic systems, evolutionary economics — are routinely used to study and describe complex behavior. All are empirically grounded; patterns are composed of smaller bits interacting with each other according to some set of rules (generally physical laws) that govern the system's evolution. The system's state at a point in time is nothing more than the summation of all the bits that compose the pattern sitting in some relationship to each other. Time ticks by. The state changes, acting under the influence of the underlying rules. And so on. Patterns emerge — they become visible — at a higher level of abstraction than the bits themselves.

Think about water molecules acting on each other under the influence of gravity and molecular dynamics to form the swirling flows of a river. Or air and water molecules interacting under the influence of the sun and Earth to form weather systems. Or neurons connected in a brain, firing in patterns dictated by synaptic nerve chemistry and sensory input. Computer nerds everywhere have seen the toy example of an animated Game of Life, where simple, iterative rules applied to a grid of dots can bring surprisingly complex patterns of swirls, gliders, and pulsars to life. Or even social networks, composed of individuals interacting over some electronic medium.

The complexity of interest to any observer is contained within the emerging patterns, not in the rules that govern the interacting layers of simpler bits. Typically, rules are simple, and behavior is complex. That behavior is unpredictable, yet it is somehow stable — that's the essence of an interesting pattern. Hurricanes are stable over a short time, somewhat predictable in their path once formed, but utterly unpredictable prior to their formation despite the (relative) simplicity of molecular dynamics. We are interested in them because they flatten houses and flood land, not because of the way air molecules bump into each other. Social networks are interesting because of how they enable flash mobs, or unexpectedly change election results, not because of the 280 characters by which individuals message each other.

The global economy is the mother of all emergent patterns, encompassing everything (including ourselves!): changing weather patterns, our brains, social networks, the ecosystem, political parties, industrial machinery, technology, satellites — all of it. The rules governing the underlying bits depend on how one defines those bits. The simplest way is similar to the Basic Picture: the "atoms" are billions of individuals interacting many times a day to form a massive economic web of activity. The "law" governing behavior is each individual acting according to an expectation of self-interest. Money is the primary medium of exchange. Transactions occur under market conditions like contract law, interest rates, and so on.

As people buy things, take a new job, invest their savings, invent a new app, or go on holiday, our emergent economy constantly reforms, ever-changing yet remaining relatively stable. Emergent patterns can be thought of as layers of varying complexity that sit atop those basic transactions and motivations — a consumer buys an iPad; the Apple Store pays its mortgage with the bank; that bank loans the money to Wall Street; an electronics factory on the other side of the globe amortizes new machinery financed by a firm on Wall Street; the energy grid feeding that factory executes a contract to buy wind power; an engineer working there invents an app to track wind flows, which increases iPad sales. Any identifiable economic activity can be characterized as a dynamic pattern emerging from those billions of transactions: flows of capital, corporate activity, brand loyalty, consumer preferences, the invention of a new technology, or the operation of large energy systems.

While the math can get heavy, a high-level understanding to inform our carbon pricing question requires picking only a few key principles these systems have in common. Complexity is open-ended (no matter how simple the rules) since there are infinite possible *arrangements* of stuff. Time is central since the rules are applied iteratively. Nonlinearity and deep complexity sit at the economy's heart as patterns emerge that create the conditions for the next ones — patterns acting on patterns. All of this implies that prediction, beyond some limited frame, is well-nigh impossible. There's not much that looks like equilibrium here. Rather, we see a highly creative dance of material, energy, people, and ideas. On the Modern View, we interpret the economy as a complex system evolving under a set of rules that are both physical and legal.

This seems like common sense. Of *course* large things are made up of smaller things; we're just describing reductionism. Of *course* things evolve over time; we see that all around us as trees grow and hurricanes blow over. Of *course* rules govern how that evolution occurs, otherwise we'd have no order to speak of (and wouldn't be here to speak of it). The critical difference from traditional economics is the conceptual priority given to time. Evolutionary economics puts time in charge, rather than giving it a back seat on the ride to equilibrium. That simple trick has huge repercussions, of which we'll focus on three: deep change, unpredictability, and path-dependency.

In the Basic Picture, change is shallow, it's a way station on the way to equilibrium. It's often treated much like an external force happening *to* the economy. The closest traditional economics gets to the Modern View may be the "gales of creative destruction" described by economist Joseph Schumpeter as a "process of industrial mutation that incessantly revolutionizes the economic structure from within, incessantly destroying the old one, incessantly creating a new one."[42] But while Schumpeter's "gales" are purportedly from within, they are *imposed upon* a model that doesn't easily capture this internal, iterative process.

On the Modern View, change runs deep, and disruption quite naturally comes from *within* without any changes to the basic model. An evolving economy acts recursively on itself, altering the very laws that govern its evolution. Think of transistors enabling the computer chip, making possible new livelihoods (programmers) and entirely new sectors like the internet and social media. Social media companies then conjure into being a new economic law — that of increasing marginal returns![xxi]

Change is so profound and deeply embedded in the Modern View that it forces a certain humility. Predicting future economic activity with this model, at any level of detail, is impossible in principle. Some high-level patterns may be robust; increased interest rates dampen capital investment, for example. But most economic activity can't be predicted. What technologies will succeed or fail? Which company's stock price will grow and to what level? Will China's debt crisis fracture global financial markets? Is solar plus storage the best new energy bet, or is it next-generation nuclear? Will batteries win the utility-scale storage fight, or will it be Hydrostor's advanced compressed-air systems, or some as yet unknown solution? The unpredictability of our economic future comes as no surprise to anyone in the investment community, and it should form the basis of our climate response.

Deep change and the ensuing unpredictability are clues as to why an economy-wide carbon price is maximally *flexible* and *efficient* in constraining GHG emissions. The trick is to see how a price on carbon maximizes the degree of freedom under which market forces operate to find solutions. The more basic the rule change, the deeper it sits in the

---

xxi Unlike physical things, like apples — where every extra one you get has diminishing value because you get sick of them — networks gain increasing value each time a new member joins. This is also known as the network effect.

economic web, the more creative potential we unlock. All the market needs is a clear signal that *identifies that constraint without the imposition of a solution*. The more basic the rule change, the better that rule can harness the very same forces that make the market deeply unpredictable; there are limitless ways to solve for the constraint imposed. Price signals are as basic as it gets since they operate at the very bottom of the economic web, affecting the self-interest of everyone, everywhere. And all actors will seek to find ways to avoid those costs (in this case, the price of carbon) by seeking ways to reduce emissions at any price lower than the one imposed.

Pricing carbon makes no decision as to *how* emissions are reduced, it only ensures that they *will* be reduced. It's a bit like gravity affecting water: we can never know what path a water molecule will take as it falls from the sky, nor the patterns that will form in a creek or river. But we do know that all water will flow downhill. Every economic actor, from individuals up to hedge funds, has maximum flexibility in deciding their own response to the rule change of a price imposed on carbon. No transaction goes untouched, from the price of consumer goods (which embed the cost of carbon in production and delivery) to the cost of project finance (which discounts the future value of a project based on expected exposure to carbon risk). Instantly, every decision from buying chocolate bars to building power plants is affected.

That sounds a lot like recent neoconservative cries of alarm: "It's a tax on *everything*!" They are correct — it is. And it must be. In that sense, the Modern View is no different from the Basic Picture — an economy-wide shift to a new equilibrium with market forces in play minimizes costs. The challenge to the neoconservatives is simple: if you truly believe markets represent the collective creative wisdom of each participant, then how else might one tap that set of resources? If you don't affect all transactions, then you limit what the market can do.

The Modern View emphasizes process over end point. We define the creative capacity of the market as its degree of freedom in solving for a constraint, which, in turn, depends on how the constraint is defined. One cannot get more basic than a price signal. All other policy choices are less flexible, because they will necessarily operate at a higher level of abstraction. Each, from regulatory change (even Jaccard's flexible regulations) to subsidies, to some extent picks winners and losers. The result is the same as the Basic Picture, but we have better tools to explain why carbon pricing works and why it can't be more efficiently replaced by other policy options.

The effectiveness of a carbon price is clearly linked to how high one sets the price: the higher it is, the stronger the resulting market signal and the more effective the policy. The higher the price, the more powerful our hose. As we've seen, this comes with a higher political cost and, if not revenue neutral, lower robustness. But other, more subtle, constraints to carbon pricing's effectiveness can be seen.

On the Modern View, the economy, like all complex systems, is highly path-dependent. What happens first defines what can — or can't — happen next. The path that got us here defines what paths we can take from here. Again, this sounds like common sense (and it is — *of course* history defines what's possible today), but the implications are far-reaching. On the optimistic side, we see new technologies open possibilities. But it goes the other way, too: history limits possibility. Grooves that reinforce existing behavior appear in the economy, just as existing streams and rivers define where water will flow after a rainfall. The prior emergence of large patterns — entrenched interests, existing infrastructure, patents, established professional behavior, or brand awareness — quashes possibilities and constrains the creative potential of the market.

The Basic Picture minimizes the process by which a new equilibrium is found. The assumption is the economy just sort of . . . gets there. A kind of abstract idealism is built in. The Modern View, in emphasizing path dependence, is more open to admitting to a multiplicity of ways in which the economy is constrained from ideal operation. These range from the mundane — existing infrastructure that blocks newcomers — to the ineffable — the psychological attitudes and political stances of entrenched interests who do the same, for example. Part and parcel of the rules by which the economy evolves from one state to another are the collective expectations we have of what will/can happen next. After all, each market participant is a human actor replete with his or her respective psychological predispositions — which include those unique to us as individuals, those we share across professions and cultures, and those we share as humans.

Think of coal plants pumping power onto the American electrical grid. Those plants form financial, political, physical, and psychological grooves in the U.S. economy that constrain the freedom of movement that can be afforded to new energy sources. Financially, those coal plants have long-term contracts that limit utilities' ability to shut them down even if cheaper sources appear. Politically, the large investors who own them exert leverage to keep them going, which includes corrupting the political and regulatory[xxii] processes. Physically, they dominate energy flows; other energy sources must respond to their presence. The nature of the grid — physical and regulatory — reflects this history: transmission lines emanate like spokes from big plants to feed the smaller distribution networks in cities. Engineers and regulators become comfortable with that operating model, and push against changes to it. Even bankers who invest in energy add to the incumbents' momentum. They much prefer investing in ancient, existing energy systems over newer ones. Absent the prior existence of coal, clean energy would accelerate into the market much faster. Our climate policies must take this path-dependency into account.

None of these actors is being irrational. They're all defending their interests as they see fit. Investors are defending their returns, politicians their backers, engineers and regulators their view of operational reliability. Path-dependency highlights a criticism from which the Basic Picture has long suffered: in order for market dynamics to unfold smoothly, one must assume each market participant is perfectly rational, has near-infinite information, and has that information in real time. None of these assumptions holds true in reality. Price signals may be instantaneous in theory, but it takes time for people to take them into account. No one person, or

---

xxii Under the Trump administration, Energy Secretary Rick Perry commissioned a study, under the guise of energy security, to establish the value to utilities of having several months' worth of energy supply stored on-site. Obviously, solar, wind, and natural gas can't do this, but nuclear and coal can. The study is disingenuous; having a giant pile of coal nearby doesn't help if the transmission system goes down — which is what normally happens in a raging hurricane.

institution, has access to full market information. But most fatally, both people and corporations act irrationally from the perspective of bare price signals, in ways that can be quite subtle — people because they're wonderfully human, and corporations because they're made up of people. Mere price signals are not enough to overcome the grooves of energy incumbents. Carbon pricing on its own will not pave the way sufficiently quickly to the low-carbon economy we so desperately need to achieve.

When I first became a venture capitalist, I found out firsthand that price alone will not necessarily motivate people to act the way you want them to, or predict they will. I thought my peers would be rational in the sense that if I showed them a great deal, with good, risk-adjusted returns, they would be motivated primarily by the deal itself. After all, that's precisely what traditional economics predicts: their self-interest is defined by the deal's potential profit. We were all going to be good little capitalist entrepreneurs. But that turned out not to be the case. Often it was clear that no matter how good the deal, they simply could not sign on because of other, softer considerations than bare naked dollars. This confused me for some time.

In one particularly telling case, we worked to overcome all objections to an investor looking at the next-generation biofuels company Woodland Biofuels (which we encountered in the last chapter). We spent days providing empirical and theoretical evidence to a Calgary-based investment banker showing our deal brought higher returns, lower risk, a bigger market, and a longer-term global advantage than the oil and gas exploration plays he normally looked at. There was no objection we couldn't overcome, which he was honest enough to admit. Eventually, he threw up his hands and declared, "Look, in all honesty, we just like oil and gas here." As goes for venture capital, so, too, for the larger private equity and pension funds. People like to stick with the tried and true, whether choosing from meals in restaurants or selecting investment priorities. Psychology matters.

While frustrating, there is more to his response than mere disinterest. The key is the difference between global versus local rationality.[xxiii]

---

xxiii The book *Freakonomics* brings this distinction to light in lots of fascinating ways.

Global rationality is the deal on a spreadsheet, all else being equal. Local rationality is the lens through which someone sees that same deal. The numbers remain the same, but all else is not equal! That's because it's rational for those who manage money to value their continued earning power over any particular deal. Their primary currency is job security, which contains all kinds of subjective (but still rational) considerations. Sticking your neck out or rocking the boat brings risk to that primary currency. Acting the same as your peers protects that currency. No one got fired for buying or selling combustible subprime paper back in 2007. But I bet a lot received flak for having turned those toxic assets down while they remained a hot commodity.

The Woodland deal brought two risks to the Calgary investor that over-rode our numbers. First, as a relative newcomer, I brought reputational risk (people love doing deals with well-known firms). Second, not only had cleantech as a sector fallen out of fashion (enough to kill the deal), but even worse, next-generation biofuels were noxious even among those investors still game for cleantech. As we saw earlier, publicly traded next-gen biofuels play KiOR had recently blown up, wiping out billions in value. That was a highly visible failure, backed by the biggest and most famous venture firms in Silicon Valley, and included the likes of Condoleezza Rice on the board. And it wasn't the only one — many others, including Coskata and Range Fuels, met the same fate backed by the same lead investor. The financial sector fled in horror from any company that looked remotely similar, no matter how different the underlying technology or economics. KiOR's failure scorched the financial earth for anyone coming later.

On the Basic Picture, one assumes that if a company can make a replacement for gasoline more cheaply than gasoline itself, the market would rationally and enthusiastically take it on. But the market is not an abstract, idealized, or rational force. The market is made up of individuals. It made sense for any individual investor to turn down a next-generation biofuels deal that could beat gasoline, even though it made no sense to the market as a whole. The currency my peers valued was a local one linked to their ongoing employment fortunes. Once I understood their local currency, it all made sense. They were being perfectly rational *from their point of view*. I am willing to admit to my own irrational tendencies; as a climate hawk, I'm sure I tend to see more value in fossil fuel busting technologies than may be warranted. That's why investment committees exist!

There is a broader lesson here. Most of the financial and corporate community is made up of good, well-meaning people who are highly motivated to act in ways that enable them to keep making a good living — even if that leads to making decisions that run counter to our collective best interests. John Kenneth Galbraith identified the tendency to reflect a group's shared ideas — even those that run counter to that group's long-term self-interest — as conventional wisdom. The example he most cited was the elite's cultural aversion to government action at the outset of the Great Depression, even though government intervention was required to save the economy and, along with it, the elite's financial self-interest.

Those well-meaning professionals whose job it is to make investment decisions regarding the trillions of dollars of capital in our money markets are not motivated to move ahead of the herd on climate risk. The psychological momentum to favor the incumbents, to repeat the same deals that made money in the past, is huge. An expression predated the internet: "No one gets fired for hiring IBM." There are lots of exceptions to this rule. But the point remains: the market is composed of human beings, all of whom are complex creatures with interests local to them. To expect large-scale market activity, including the actions of large funds and energy companies, to be impervious to these myriad local interests is naïve. Markets are not like an idealized video game. They are more like a real-life sports game. The human element dominates.

A price on carbon will eventually overturn these subtle barriers that inform path-dependency. But we again face the question of efficiency versus efficacy: if your house in on fire, you'll trade some efficiency for speed. Because entrenched interests and systems constrain new entrants, they quash the very market dynamics we are trying to unlock. By being honest about the multiple ways in which history constrains market behavior, we have an argument to go beyond carbon pricing. There are lots of ways to short circuit these market barriers. Better information flow that affect the norms of the investment community is one of them, in an attempt to change the way they perceive carbon risk specifically, and climate disruption in general. Making climate solutions "trendy" may sound a bit trite, but it's one way to short circuit very real psychological grooves that prevent mainstream investors from taking a more proactive stance on climate.

In the Modern View we have a framework in which carbon pricing is positioned as a singularly pragmatic, though imperfect, response. It's the only policy tool that does two seemingly paradoxical things at once: it tames the market by tamping down high-carbon activities while simultaneously leveraging its creative potential to unlock low-carbon activity. It has been stripped of political ideology. For the right, the market is champion as its creative potential is unleashed. For the left, market forces are tamed, harnessed to pull us to a more sustainable path. There is nothing more centrist, on the Modern View, than pricing carbon — a critical foundation of any Climate Capitalism.

Our understanding of any system is enabled and limited by the language we use to describe it. There are heaps of math to dynamic systems, and we can perhaps understand them a little better by wringing out a final metaphor. Those on the right tend to view markets like a jungle, a natural system subject to natural law. On this view, it's best left alone to grow, wild and untamed. The left sees markets akin to an artifact, engineered and designed by humans. Here, we imagine some piece of infrastructure, like a factory or car, and it's by imposing good design that the machine functions well. The truth is somewhere in between.

Imagine the economy as a modern farm. The crops and trees grow according to their own internal dynamics, akin to market forces. We don't pretend to be able to replace or replicate those complex and effective dynamics to see a yield from what we have planted. But for the farm to be fully productive, we add wires and posts to guide growth where we want it (like a carbon price), we prune to trim excess (like regulations to sweep away vested interests), we add fertilizers to maximize nutrient uptake (like funding R&D and accelerating technology to market), etc. It's in that balance of leveraging natural forces and guiding them to our purpose that we find the most fruitful view of our economy. And how we best approach building the scaffolding of Climate Capitalism.

. . . To the Absurd

Despite the pragmatic, non-ideological nature of carbon pricing (in even a Chicago School–based view), traditional defenders of market forces in North America — Conservatives in Canada and the GOP in the U.S. — have been busy grinding their climate grist on the carbon pricing mill. Previous to them, all policies were equally deserving of scorn since the problem wasn't real. Now that climate denial is limited to loons or charlatans, one would be forgiven to expect the political right's preferred solution to be some kind of carbon price. After all, it puts those hallowed market forces in the center of the action. Alas, that's not the case. Conservatives and free-market think tanks across North America, from provincial and federal political leaders to the Fraser Institute, deride the "job-killing carbon tax."

That the most mainstream of market mechanisms is controversial to those on the political right — who, at least in Canada, now publicly agree we have some kind of a climate problem — can be evidence of only two things: either their public declarations of concern over climate are disingenuous, or they are willing to sacrifice their own ideological foundation for short-term political opportunism. Or both.

Conservatives have long championed market-based solutions to any given problem, while liberals typically lean toward a more regulatory- or government-centric approach. Over the past couple of decades, as the left fought to create support for climate action, they increasingly endorsed carbon pricing as their policy of choice. The idea was simple: a policy with support across the political spectrum would not only increase their own political fortunes, but put the climate debate safely outside normal partisan conflict. As evidence, look no further than Alberta, where longtime governing Conservatives and upstart NDP both enthusiastically supported a carbon price (albeit before Jason Kenney arrived on the scene). In Canada's most carbon-heavy economy, no less.

Like military spending in the U.S. or health care in Canada, the thinking went, the political debate about climate would then shift to arguments about details: how high a price, what to do with the revenues, how to deal with cross-border economic competitiveness, what to do in addition to carbon pricing, etc. By taking on a traditionally conservative policy solution, the left was working to ensure that climate action would become another "third rail" of politics — something no one can touch. Particularly if the carbon price is revenue neutral; it is freed from right-wing attacks as just another lefty attempt at "big government." A revenue-neutral carbon price is as centrist as it gets!

But now we're in the world of crazy, a world of factions impervious to sensible solutions and unwilling to collaborate with those on the outside. Witness arch-conservatives like federal leader Andrew Scheer, newly minted Albertan Premier Jason Kenney, and Ontario's Premier Doug Ford, all fighting tooth and nail against any form of carbon pricing, revenue neutral or not. Each of these politicians has (tepidly) acknowledged the climate problem, but none has offered alternative solutions.

Premier Ford now holds office. It's telling how little he has to say about climate plans going forward. He, in particular, had a golden opportunity to play to his base and simultaneously own the climate file. He could have rendered the existing carbon pricing scheme revenue neutral by returning the money to voters. He'd get all the political gains associated with his name on checks in the mail every quarter, and he'd position his party as a climate champion. Instead, he's ditched the carbon price entirely, and is pouring tax dollars into fighting the federal government in court over their right to replace it.

At risk is not just Ontario's climate and economic credibility (industry is already reeling and investment vacillating[xxiv] from his reckless repudiation of cap-and-trade). Politically, his decision makes no sense. When the feds replace what he ditched, it will be the federal government who decides where the money goes. It will be the federal Liberals who will send out those checks to taxpayers. By picking a fight with the

---

xxiv GM was a strong supporter of cap-and-trade. While their recent announcement to close its plants in Oshawa can't be attributed entirely to Doug Ford's decision, it was likely a factor. GM is fully committed to moving aggressively into electric vehicles, and surely had an eye on some of the billions of dollars raised by cap-and-trade to help offset those development costs.

feds over carbon pricing, he's doubled down on his entrenched position, which is a dead end for conservatives.

Things get messy when political expediency gets in the way of economic efficiency. Imagine the following scenario: I operate a chemical plant in Ontario. That plant is a best-in-class, energy-efficient operation compared to my competitor in Mexico, because of the historical costs of energy. Emission reductions are therefore cheaper in the Mexican plant than my own, because the easy investments in energy efficiency haven't yet been made there. A regulatory approach requires me to reduce my own emissions. Cap-and-trade means I can choose to pay for the cheaper reductions in my competitor's factory. While it's lousy optics to see local money flow offshore, the irony is that may be the best way to keep Ontario industries humming while other jurisdictions catch up.

So why are conservative politicians rejecting proven tools like cap-and-trade and carbon taxes when price signals are the bread and butter of conservative politics? Ford, Kenney, Scheer, and other conservative politicians who dismiss revenue-neutral carbon pricing are now stuck with a regulatory approach — if they are to act on climate at all. That's the only option outside pricing carbon. While potentially effective if aggressive enough, regulations are anathema to conservative political roots. And regulations aren't free, even if the costs aren't as visible. Perhaps conservative-leaning governments will put in place an effective and efficient regulatory environment. But I doubt it — nothing stinks of big government like regulations. And nothing cozies up to Bay Street and Wall Street like the very market mechanisms they're tossing aside.

The only way to explain why conservatives across the country have thrown away their own best policy tool in favor of a highly visible fight with the federal government is short-sighted political expediency. Making carbon pricing a wedge issue is a nasty partisan play for political advantage. It certainly plays well to the extreme-right ideologues in the party. But I believe that strategy will fail when it's pitched to the general public. A majority of Canadians in every political party want action on climate. Braying against carbon pricing may play well inside party politics — like the primaries — but it's not something the electorate will endorse. We'll see by the time this book is published. Canada's federal election in 2019 will be fought largely over this very issue. (Although recent SNC-Lavalin controversies may override all else. It would be a real shame to lose a national carbon price just because someone felt the need to cuddle SNC!)

In retrospect, it may have been better for everyone if the left had stuck to its guns and pushed a regulatory approach rather than endorse the policy mechanism most favored by the right. That would have left room for the right to champion their own approach (carbon pricing) instead of using that same tool as a political piñata upon which their base can vent frustration.

It wasn't unreasonable to assume that someday leaders of all parties would act like adults on climate risk. Carbon pricing anticipated that day. Perhaps when the political dust settles, we'll get a revenue-neutral carbon price endorsed by all parties — a policy that they can all prioritize and agree on, similar to basic health care in Canada or military spending in the U.S. Parties can then differentiate themselves by what *else* to do about climate. Because carbon pricing may be a necessary part of our climate response, but it's no longer sufficient.

The central challenge is we're so late to the game that a carbon price in isolation must be so high as to be politically impossible — even in more rational times. A recent IPCC report detailing efforts required to stop warming at 1.5°C put the global price at more than $400 per ton, and for 2°C nearly $100 per ton. These numbers are above anything in the world today, with the possible exception of Sweden. The IPCC's prices are *predictive*: set the price to x and we can expect the market to do y. A slightly different approach based on *fairness* sets the price equal to the economic damage from each incremental ton of carbon (the "social cost of carbon"). That approach, while intellectually sound, relies on unworkable[xxv] economic models to come up with a number.

Independent of all these approaches, the practical limit is what the public will accept. I doubt that will ever be high enough to stave off

---

xxv See *Waking the Frog* for a comprehensive critique of the leading models of this kind, such as DICE (Dynamic Integrated Climate-Economy model) and RICE (Regional Integrated Climate-Economy Model).

catastrophic climate disruption. As a practical matter, we will need more policy options on the table than a carbon price alone, to which we soon turn. But before that, let's consider the most powerful allies an environmentalist could have: the corporate sector. Normally, these two groups are at odds (and seen as eternal enemies by Klein and the far left), but Climate Capitalism makes the case that we can — and *must* — find a way to recruit the world's most powerful firms and investors into the climate fight. Without them, I fear, the game is up.

CHAPTER FOUR

WHITE HATS AND BLACK HATS: GETTING CORPORATE LEADERS IN THE FIGHT

In traditional TV Westerns, it's easy to tell the goodies from the baddies: good guys wear white hats and bad guys wear black ones. Just as TV wipes away moral ambiguity in order to present an easy-to-digest, oversimplified moral landscape, so, too, the corporate universe is often presented in stark terms. The White Hats — like Elon Musk's Tesla and SolarCity — work on the side of angels to bring clean, everlasting light to the masses. The Black Hats — ExxonMobil as lead villain with the Koch brothers in a supporting role — manipulate public opinion, bribe politicians, and work to keep fossil fuels' stranglehold over energy systems.

There's some truth to these caricatures. Elon Musk is indeed motivated to solve the climate crisis. ExxonMobil and Koch Industries (among others) have (and still do) fund anti-science, climate-skeptical think tanks and political action groups.[i] Many of their fossil fuel peers work to block cleantech competition and lobby against environmental regulations, including any sort of carbon price. And the predatory,

---

i   For a great overview of the corporate climate Black Hats, see Naomi Oreskes's *The Merchants of Doubt*.

near-sociopathic attitude of some in the financial sector is well-documented[ii] and remains a particular source of bile for many post-bank bailouts in 2008–2009. But reality is far more complicated than these examples might imply.

Certainly, there are good and bad corporate actors — just as there are nice and nasty celebrities, honest and dishonest accountants, and likable and distasteful athletes. But the vast majority of companies and investors — as in all human categories — play somewhere in the middle. Certainly, all are motivated to a greater or lesser extent by self-interest, and many are prone to willfully ignore what appears most inconvenient to that self-interest. But broadly speaking, corporate leaders are equally considerate of human decency, have real concern for the future, and respect traditional norms of truth-telling and rational debate when it comes to important risks. And they generally provide consumers with the goods and services they want. If they didn't, they'd go out of business.

Is Facebook a White Hat or Black Hat? Walmart? It's not hard to portray these two global giants as evil. Facebook's notoriously incautious approach to individual privacy, willful blindness to being agents of (and benefiting from) nefarious political actors with bad intent, to say nothing of their equally troubling relationship with partners in the health sector[iii] — was enough to convince me to delete my Facebook profile in protest. After due consideration, the U.K. government recently portrayed them as "digital gangsters." And Walmart has ravaged local economies and communities by leveraging their buying power to strong-arm thousands of retail outlets out of business. And, of course, their treatment of employees is not top-tier.

---

ii   Matt Taibbi has done a wonderful job documenting the nastier and more destructive nature of Wall Street.

iii   While the political scandal involving Cambridge Analytica was breaking in early 2018, Facebook was deep in discussions with hospitals and other medical groups to combine their own data with anonymized medical profiles. By "hashing" that data — reverse engineering the identity of the supposedly anonymized medical record using extensive data available inside Facebook — one can imagine the value to the medical insurance industry.

Yet Facebook provides huge demand for clean energy as they look to power all their data centers with low-carbon energy sources. Those efforts are real, evidenced by their meeting a short-term target of 50 percent clean energy by 2018. New data centers come online matched with a power purchase agreement (PPA) to offset that energy load with renewables from the grid. Apple, Google, and others do the same. And Walmart doesn't just lower the cost of underwear, socks, and televisions. They're fast becoming a cleantech star[iv] themselves, both within their own operations and by leveraging their supply chains to force suppliers into making better choices about their own energy use — just as they use that same leverage to squeeze profits.

It's tempting to divide the corporate world into White Hats and Black Hats, much like we refer to the left and the right on the political spectrum, but aside from the obvious and outlying cases — like Big Tobacco's outright lies to Congress about health effects, Exxon's repression of public debate on climate risk, or Shell's trampling of human rights in Nigeria — it's not that simple. Most corporations operate as Gray Hats, neither evil nor good, merely pursuing profit within the law and generally abiding by normal bounds of human decency.

A more interesting question is: how might we turn those many Gray Hats white? The vast majority of corporate leaders view neither themselves nor their organizations as moral agents in the climate fight, but as mere actors in a larger economic scene. What sort of nudge might enable them to act in concert with a faster transition to a low-carbon economy to better align that neutrality with a good outcome in our shared climate fight? Getting corporate leaders and their organizations onside to fight for a stable climate, independent of what policies may *force* them to do, is crucial for a robust, proactive Climate Capitalism. Attitudes matter!

---

iv   That global shipping underpins their business model is a harder nut to crack, but taking advantage of global supply chains is hardly unique to Walmart.

THE ENERGY GIANTS

Looking to attitudes adopted by various actors in the traditional energy sector is instructive as we seek to enlist corporate leaders in the climate fight. Aside from the malfeasance demonstrated by Black Hats like ExxonMobil, we might divide the sector's role in climate action into three categories: forced actors, those who view emission constraints as a legal burden by which they must abide; willing actors, those who dabble in clean energy independent of their legal obligations; and leading actors, those who enthusiastically push against the normal limits of corporate constraints to become an active force for positive change across both technology and policy.

Forced actors form the vast majority of energy giants. Most (now) acknowledge climate risk and view the transition to a low-carbon economy as a constraint placed on their normal mode of business — a business that will nevertheless continue to thrive for the foreseeable future. Investments in clean energy are driven by a sense of regulatory compliance rather than strategic advantage. Cleantech is applied behind-the-fence (within existing operations) to reduce the pollutants associated with production of their primary product — fossil fuels — in order to comply with regulatory pressure. They work to green their product under the assumption that demand for it will not falter, rather than finding or developing alternative sources of energy.

Cenovus is a typical forced actor. A stellar performer in Canada's heavy-oil patch, their two Alberta plants are squeaky clean with little to no above-ground toxins. By melting bitumen below ground (in situ) instead of mining the stuff in an open pit, they produce no tailings ponds. But it takes a massive amount of natural gas[v] to convert that bitumen to usable oil. First, they burn enormous quantities[vi] of the stuff to melt the

---

v   It's not a stretch to say that much of the heavy-oil patch is trading a carbon-light fuel (natural gas) for a carbon-heavy fuel (heavy oil), with modest gains in total energy content (about 4:1 compared to upwards of 20:1 for conventional oil).

vi I've seen the natural gas burners at a Cenovus operation, and they are truly impressive — a long row of rocket-scale blue flames.

bitumen. Then, they strip hydrogen from yet more natural gas to add to the heavy oil to make it resemble conventional oil. Their focus is to reduce that energy intensity — the amount of natural gas required per barrel of oil. Cenovus is as well-meaning a heavy-oil company as you'll find, taking a lead role in defining best practices within their industry and supportive of Alberta's carbon-pricing policies. Hardly the devil, but they're incapable of championing a low-carbon future; the asset they're sitting on precludes them taking on that role.

Willing actors go further than established best practices. They contribute to our understanding of climate risk and engage in research and dialogue about what sorts of energy systems are needed to keep global temperatures down. Many embrace the transition away from coal and heavy oil to natural gas as a primary response to emission constraints, but also make investments in non-fossil energy production like wind, solar, or biofuels. This can sometimes be a bit of a show, but is often a genuine effort to explore ways to diversify their business. If they push the bounds of technology, it's generally to support the development of carbon capture and storage.

TransAlta and Shell are typical willing actors. Once fully committed to coal, TransAlta began dabbling in wind after being prodded by their sustainability officer, Bob Page. They bought Vision Quest, their first wind farm, from Canadian wind pioneer Fred Gallagher in 2002 for just under $40 million. At the time, the business community saw the deal as largely immaterial, although a couple of analysts[vii] at UBS Warburg wrote, "There is mildly positive strategic significance." How right they were! Seeing decent returns, TransAlta stuck with it, deploying hundreds of millions annually into expanding that "immaterial" side of their business.

Eventually, they spun off their renewables division into a separate, publicly traded company. Beginning in late 2014, while investors watched in horror as the value of TransAlta's coal assets crashed, their spirits (and equity) were lifted by the 70 percent ownership TransAlta had retained in that spin-off — the value of which kept rising. At one point, that share of ownership accounted for the entire value of the enterprise! What began

---

vii   Andrew Kuske and Ronald Barone.

as an experiment came to form the basis of substantial, long-term value. It will be instructive to see how much that dynamic informs their peers going forward, as heavy oil goes the way of coal.

Shell has long engaged in serious dialogue on climate. In 2016, I enjoyed a robust debate in Alberta with their chief climate change advisor, David Hone. He is a leading and well-informed voice on the long-term difficulty of building energy systems capable of keeping temperatures down and the economy humming. Hone is backed by a team of researchers at Shell who conjure up scenarios, ranging from the bleak (Mountains) to the hopeful (Sky). While Sky marks our best efforts to achieve what was promised in the Paris Agreement, Mountains is described as follows:

> This is the world with status quo power locked in and held tightly by the currently influential. Stability is the highest prize: those at the top align their interests to unlock resources steadily and cautiously, not solely dictated by immediate market forces. The resulting rigidity within the system dampens economic dynamism and stifles social mobility.

This is hardly the work of a company whose climate ambition is merely to comply with a regulatory burden. It's a frank acknowledgment of the deep, divisive nature of climate ambitions. Shell has gone so far as to break an unspoken taboo in which the large and powerful don't mention their own possible mendacity in protecting what makes them most comfortable.[viii]

The primary drivers of Sky, their most aggressive scenario, are a combination of natural gas and bio-energy carbon capture and storage (BECSS). BECSS is like CCS except you burn biomass, which is then regrown. The result is negative-carbon energy, since you're constantly

---

viii For fans of John Kenneth Galbraith, the iconoclastic economist would be pleasantly surprised to see an institution, whose executives are surely deeply embedded in the Culture of Contentment, working nevertheless to change the conventional wisdom protecting that sense of contentment.

capturing, burning, and burying carbon from the air. The Sky scenario is aligned with the ongoing health of Shell's core business, with tweaks. From their perspective, that's just being realistic.

There is a flaw in their reasoning. In preparing for the debate with Hone, I dug into these scenarios. I found them refreshingly rigorous and accurate. Tellingly, though, they treat the technical and economic progress of alternatives like solar and storage as external variables to their model. In Shell's world, others are responsible for generating the innovation that increases the economic viability of alternatives. Shell acts as a passive recipient of those innovations. In other words, they assume no role or responsibility in improving the odds that non-fossil energy plays a bigger role. Which, in turn, speaks to the need for natural gas and BECSS. It's a neat trick. But it's just a model.

I've seen a similar stance adopted by lots of energy executives who, when asked about the incompatibility of climate constraints and their own long-term economic health, point to charts showing growing global demand for fossil fuels. "What can I do?" they shrug. "We're just filling the needs of our customers." Adopting such a passive role in energy markets always struck me as unconvincing, a false humility. These are captains of industry, for whom risk-taking and market domination are second nature, who bend the physical world to their will as they execute ever-more-dramatic (and frankly incredible) feats of engineering to chase dwindling supplies of oil deep underground in harsh conditions. If you can drill miles down in the deep Arctic to pierce a pocket of oil, you are surely capable of bending that energy demand curve a little.

This passive attitude to market demand is the defining difference between a willing actor and a leading actor. Leading actors believe they can change global energy demand by building alternatives to fossil fuels. For them, that demand curve isn't immovable. A fossil fuel company that actively works to undermine the long-term market for their own product is a leading actor in our climate fight. Innovation in real alternatives is not some external variable for them. It's an internal drive, a way to define their corporate ambitions and motivate their engineering and finance teams. Innovation is a potential source of long-term strategic advantage against their slower-moving, more passive peers. Long live the leading actors!

Compare North America's willing actor TransAlta to Ørsted, a Danish peer. Back in 2010, more than 70 percent of Ørsted's power plants burned fossil fuels. In 2013, they announced they would divest from these and move into renewables. They proved to be serious, and moved aggressively into offshore wind, acquiring and developing assets around the world. Today, that 70 percent share comes from renewables. And they're not done. By 2020, less than one-tenth of their fleet will be carbon-emitting. Best of all, their market capitalization is up by two-thirds since their IPO in 2016, a marked improvement over their peer group.[43]

Second only to Ørsted is Equinor (formerly Statoil). The Norwegian oil and gas giant changed their name to reflect their decision to become a new kind of energy company. I first encountered Statoil in the mid-2000s when I attended a cleantech conference in Stavanger, Norway. At the time, I was weary of North American corporate executives endlessly debating the existence of climate disruption, never mind its relevance to their industry. Yet in Stavanger, I witnessed the CEO of one of the world's largest fossil fuel companies excitedly explain the engineering challenge of developing massive, floating offshore wind turbines. At the end of his talk, I asked timorously, "You're an oil and gas company. Why are you taking on the engineering risk of offshore wind?" I was dumbfounded by his answer, delivered in a gentle Norwegian lilt: "Because it is the right thing to do. And because we are able to do it!"

And they did. Just as in the artist's rendering in a video I saw a decade ago in Stavanger, Equinor developed Hywind, a floating platform capable of supporting massive wind turbines twice the height of the Statue of Liberty. Instead of deploying existing wind technology (like TransAlta), they committed capital and expertise to invent a better one. The world's first offshore floating wind farm off the coast of Scotland produces more wind more of the time[ix] than conventional wind farms, because over deep

---

ix   Hywind's capacity factor — the proportion of energy actually produced compared to what's theoretically possible given 24/7 of perfect wind conditions — during the first three months of full operation was an astonishing 65 percent, compared to somewhere between 30 to 45 percent for a typical windfarm.

water is where the best winds are found. They recently added giant battery packs to store that energy and deliver it when it's needed most.

The key point is not that Equinor built a wind farm. Lots of companies do that. It is their commitment to changing the economics of wind, to use their own muscle to make wind more competitive against oil and gas in a country where electric vehicles pose a profound threat to oil demand. When asked why they have chosen to play a role at the cutting edge of renewables, their answer is disarmingly simple: "No other company combines our decades of offshore experience with our project execution capabilities."[44] In other words, because they're an existing energy giant, they're in the best position to be a new kind of energy giant.

Like their less progressive peers in the energy sector, Equinor also continues to invest in CCS and to lower emissions in their legacy business, but the difference is that they actively work to undermine that core business, even as they continue to operate it. That's what makes Equinor a leading actor. Their New Energy Ventures Fund is putting $200 million to work across the low-carbon energy spectrum.

One way to gauge where an energy company sits on this spectrum is to look beyond the kinds of technology they deploy (like wind and solar farms) to what kinds of bets they place on technology change and how big they are. That's what drives us past scenarios like Mountains and Sky. While realistic, they're limited in their imaginative capacity to what already exists. Those who invent the future are those who create it. Whatever lies beyond Shell's vision is that which we must seek. That's what Climate Capitalism is meant to create. And that takes money.

Venture capital is the lifeblood of innovation. It's a fickle sector, often chasing the latest fad, from Twitter to FinTech and back. As we've already learned, cleantech was in favor in the early 2000s, but when Silicon Valley lost its shirt, funding for the sector dried up. One might think the fossil fuel giants would step in, given their access not just to capital, but to the very markets and infrastructure that cleantech seeks to disrupt. Their ability to scale technologies past the point where traditional venture can get them — large, industrial-scale facilities — speaks to a real competitive advantage. And, given that cleantech is a potential threat to their businesses, they should be motivated to get into the cleantech venture game even as a defensive measure.

Yet they are largely absent. In preparing for the debate with David Hone, I was shocked to find out just how absent. Little ArcTern Ventures, my own fund, had by that time invested only around $50 million over five years into companies pursuing zero-carbon or near-zero-carbon energy technology. Shell, a massive organization for whom that kind of money is a rounding error — and who openly engage on climate risk and speak about their cleantech commitment — had at that point invested about the same![x] Small wonder advancements in solar or storage are treated as external variables in their scenarios; that view reflects the facts.

In a promising sign, Shell now appears to be following Equinor's lead. Their own new group, Shell New Energies, is getting more aggressive in their approach to cleantech investing as part of a well-capitalized effort within the company to begin the long transition to a low-carbon power company. They've even quietly purchased some electrical utilities to get their feet wet in what's a very new space for them. In 2020, they're planning to spend $1.5 billion, or 5 percent of their capital budget, to develop that electricity business. That compares well to Equinor at 6 percent. What's driving this changed attitude? A fear that the days of gasoline are numbered, driven by faster-than-expected development and adoption of electric transport. When asked about the U.K.'s plan to ban sales of new fossil fuel powered cars by 2040, Shell's CEO Ben van Beurden said recently, "If you would bring it forward, obviously that would be welcome."[45] That's an oil executive talking!

But committing even less than a tenth of capital budgets to renewables triggers pushback. The large investors who back those companies are skeptical that clean energy generates the same returns as traditional stuff and whether management team skills and corporate cultures are transferable. Ørsted's and Equinor's experience should blunt those

---

x   Excluding carbon capture and storage (CCS), which may be well-meaning, but is really a bet on continued business-as-usual. As long as the industry can point to their work on a potential long-term solution for their own emissions, they reduce the pressure to find real alternatives.

concerns. For now, the broader industry remains entrenched as forced actors. Its investment in cleantech remains insignificant compared to the $170 billion spent annually to find new reserves. It also pales next to the billions spent on CCS. That comparison — billions versus millions — is evidence of the industry's lack of drive to create truly transformative technologies. If this seems unfair, the Canadian cleantech industry — a bunch of small enterprises treated as an afterthought within the larger business community — spends more on R&D than the entire oil and gas sector combined. If you want to know wherein lies the future, look to R&D spending.

This is not to accuse the large oil and gas companies of inaction. But the big bucks still find their way to support the business of selling fuels extracted from the ground. Cenovus had a venture fund that invested in "real" cleantech. But they've backed off that more progressive approach. Cenovus's new fund, Evok Innovations — a $100 million (over ten years) co-investment with Suncor Energy — marks a return to the usual focus on behind-the-fence technologies: lower pollutants in tailings ponds; suck more oil and gas from an existing field; make processing operations more efficient; move heavy oil through pipes more easily. These may be good things for Cenovus, but they're far from transformative.

It's good to see Shell (and others) acknowledge the problem, but it's not enough. David Hone does a great job making the public aware of the company's view of the climate problem (it's real), its level of difficulty (very), and the various scenarios they've mapped that might limit warming. But the bottom line is most of the oil patch is *reacting* to energy innovation instead of *creating* it. That's changing, but too slowly to see the industry as real climate partners. Perhaps Equinor can take the longer view because they're majority-owned by the Norwegian government. If that's the case, it's an argument to renationalize the energy giants if they can't see their way to stepping up more aggressively.

The larger lesson is we shouldn't expect traditional energy companies to drive the transformation we need. Their job remains to deliver what the market demands. But if the cleantech industry can change what kind of energy faces demand and disrupt traditional energy flows, then there are willing partners in the traditional energy space that can bring scale. The trick is to focus on those companies that have made it the mandate of their executive teams to seek out those opportunities.

So, how do we turn those Gray Hats white? From a cleantech entrepreneur's perspective, you can't do it directly. Nothing I say or do will convince ExxonMobil to change their view or get BP to invest in solar again. But we can support the White Hats like Equinor, ensuring their commitment to low-carbon energy is rewarded in the market. And that, in turn, will push Gray Hats to white as they see their more progressive peers gain advantage. More and more cleantech partners will come from the energy patch, of that I'm certain.

But the single biggest barrier in changing the behavior of Gray Hats is not entrenched attitudes or a lack of viable alternatives. Even a clear-minded CEO who decides to support wind, solar, and storage can't do so at the expense of walking away from the giant pile of coal or oil she's sitting on. That's leaving money on the table. Her board of directors has a fiduciary duty to maximize its value to investors. That means digging it all up to be sold and burned. While they may be persuaded by a decent argument about the long-term advantage to support alternatives, they will not do so at the expense of leaving fossil fuel assets in the ground.

We should expect the vast majority of energy companies to fight to the bitter end for their right to dig up and burn all of the reserves they carry on their balance sheet, regardless of whether they're Black, Gray, or even White Hats. Those executives are not evil people. They're just playing by the rules in a very well-defined game. What's called the carbon bubble is a clear indication of the fight we have on our hands, and that's why it's the investment community who hold the real keys to change.

CARBON BUBBLES, OR, THE GREAT FOSSIL FUEL STOCK BUYBACK

Bubbles have been around ever since people bought and traded stuff for profit. And they cover a huge variety of assets: stocks in the East India Trading Company, tulip bulbs in Holland back in the seventeenth century, NASDAQ tech stocks of the late 1990s, and most recently,[xi] American real estate. The value of an asset rises far beyond what would normally be rationalized, driven by a kind of hysterical optimism as people come to believe the price of something will rise because other people believe it will rise because more people believe that, too. And so on. It's like a legal Ponzi scheme, driven by the greater fool theory.[xii] Those inflated values can deflate very quickly, literally on a whim, when people's minds suddenly switch gear. All bubbles eventually pop.

Carbon Tracker, a group out of the U.K., makes a credible argument that the stock prices of those traditional energy companies are way into bubble territory (the blacker their hat, the bigger the bubble). In this case, it's not because of hysterically optimistic expectations about the future value of those stocks being driven ever higher by ever greater fools, but because their value is unsustainable under even the more conservative assessments of climate risk.

A fossil fuel company's value is roughly equal to the net present value (NPV) of their proven reserves.[xiii] Proven reserves are coal, oil, and gas in the ground over which a company has ownership or access rights, and which can be extracted economically (assuming some market price). It's what is secured in the cupboard, ready to be sold on the market in the future. In financial terms, those reserves sit on their balance sheet. To maintain their share price, they need to replace what they sell. That's why companies work so hard exploring for new reserves, from the bitter

---

xi   Arguably, as I write this paragraph, lots of other assets — from stocks to Chinese corporate bonds, and even real estate — have moved dangerously close to bubble territory. This is in reaction to the massive amounts of cash generated by the Federal Reserve's quantitative easing (QE) combined with ultra-low rates of interest put in place as a reaction to the U.S. housing crash. Always seeking places to maximize returns, that cash has to go somewhere, and it's been slowly driving up asset prices across the globe.

xii   I may be a fool to buy a tulip for more than my annual income, but there will be a greater fool coming along soon enough to take it off my hands!

xiii   According to OPEC, "an estimated quantity of all hydrocarbons statistically defined as crude oil or natural gas, which geological and engineering data demonstrate with reasonable certainty to be recoverable in future years from known reservoirs under existing economic and operating conditions."

Arctic to the bottom of the Gulf of Mexico. They need to top up what's in their cupboard. That's the basis on which they're valued.

Here's what the market is telling us today: since the total value of the fossil fuel sector is roughly equal to the NPV of the total proven reserves, the market participants are assuming all those reserves will be sold. An energy company's stock price isn't based on what they sell in a given year, but what they've got in the ground. The greater the risk those reserves[xiv] won't be sold, for any reason, the higher the discount rate investors would apply to get that NPV and stock price. Which is a fancy way of saying share prices are meant to reflect the odds of selling the stuff in the cupboard. And investors today are betting, as they have for decades, that *all* the stuff in those companies' cupboards will be sold.

Why is that a bubble? This is where the work of Carbon Tracker comes in. They made two simple[xv] calculations. First, how much carbon is in those proven reserves that make up the balance sheets of energy companies? The answer is around $27 trillion worth. Second, how much total carbon can be emitted to give us a fifty-fifty chance of hitting the Paris target of 2°C? That's about $7 trillion — our total allowable carbon budget. To succeed in our stated international commitments, we can burn only *one-quarter* of what's sitting in the collective cupboard of energy companies. The 2°C target is irreconcilable with existing energy company valuations — there's a $20 trillion gap!

There are two ways to read that gap. For those who believe markets reflect the collective wisdom of millions of investors, and are normally correct in assigning value, then the market is telling us we will fail miserably on climate. It is saying, loud and clear, we'll sail past 2°C of warming

---

xiv   Not all reserves are created equal: the cost of extraction links NPV to long-term fluctuation in prices; a high degree of technical and environmental risk (e.g., the high Arctic) leads to higher discounts; carbon content per unit of energy (coal and heavy oil vs. natural gas) links discounts to policy risk.

xv   They are simple on the surface, but really complex to actually calculate, since companies and countries tend to be a bit opaque here, Saudi Arabia in particular. They tend to hide stuff — like the real amount of their reserves, how economical different fields are to bring to market, etc.

well into the mind-bending horror of 4°C or even 6°C. The industry, on this view, is fully confident that government regulation and cleantech competition will not put a dent in demand for their product for the foreseeable future. Even as the planet becomes more and more unlivable!

For those who believe markets can be irrational for long periods of time, and that sooner or later, we will get serious about addressing climate risk, the market is saying something very different: those stock prices must eventually come crashing down. If we hold warming down to 2°C — for any reason, be it regulatory constraints, competitive clean energy, or (most likely) a combination of both — we will not burn more than a quarter of those proven reserves. Even if we were to stop at 2.5°C, or 3°C, the vast majority of those reserves will remain in the ground as stranded assets. Which means the stocks are way overvalued[xvi] and, hence, this is a bubble.[xvii]

You can't have it both ways. The math just doesn't work. One cannot simultaneously believe the fossil fuel industry is correctly valued and international efforts to contain global warming will succeed, even moderately. Both options look bad. The former for the financial markets, the latter for everything (including the financial markets). Only one of these risks is manageable.

Mark Carney, governor of the Bank of England, worries a lot about the carbon bubble. Twenty trillion is a lot of money, even for central bankers (who can otherwise be somewhat blasé about the t-word). Twenty trillion is three times the size of the U.S. home equity bubble of 2009. Carney sees the fallout from a collapse of energy asset prices, and write-down of stranded assets, as the single greatest threat to long-term financial stability. When one of the most influential members of

---

xvi   Not quite by a factor of four, due to the role discount rate plays in calculating present stock values, but something close.

xvii There is another way to see the bubble. The global economy will not function up near 4°C–6°C of warming. Long before then, we're into anthropocentric nightmares, which diminishes the value of those reserves, which, in turn, means the stocks are overvalued. But then, so, too, are all stocks that rely on a healthy, functioning global economy. See Part Three: "Welcome to the Anthropocene."

one of the most closed and conservative clubs in the world talks of the "catastrophic impact" climate change will have on financial markets, it's surely time to worry.

The demand we must make of the fossil fuel industry and its investors is as audacious as it is mathematically inevitable: they need to leave $20 trillion of assets in the ground in order to mitigate our collective climate risk. The last time in history it was demanded of some portion of the economy that they leave that kind of money on the table was when the United States banned slavery. While there is no moral equivalency between slave-holding and burning fossil fuels, there is an economic one: estimates[46] in today's money of the economic value of slave-holdings as one of the drivers of the U.S. economy is on the order of $10 trillion. The U.S. fought a four-year bloody civil war over the issue.

Carney believes the way to engineer a soft landing, to gently deflate the bubble, lies in transparency. If investors had better data about the carbon (and climate) risk in their portfolios, they could better react to those risks. This seems self-evident. As more investors see those risks, quantified in a way that is both transparent and a shared foundation for assessment, they will react. Healthy competition between investment professionals to manage that risk will do two things. First, it will put more pressure on energy companies to change. Second, it will both cause and manage the rush to the exits as smarter, more progressive investors begin to move first, providing cover to their slower-moving peers.

The good folks at Carbon Tracker are working to provide that transparency. They've done lots of nuanced calculations that reveal different risks: by geography, fuel type, at the corporate level, for different temperature targets, and so on. They estimate the value of stranded assets — operating energy plants and proven reserves — that need to be turned off before the end of their useful life to hit various temperature targets. They calculate the amount of new investment by fuel type, upstream (production) and downstream (processing), the market can absorb. Carbon Tracker is telling investors who support the world's largest industry, in as quantifiable a language as possible, your investments are at risk, the actions of the companies in which you've invested are incompatible with climate commitments, and sooner or later huge financial losses will come your way if you don't wake up soon.

The related divestment movement is an effort to convince investors to sell off their high-carbon energy holdings. The goal of this ad hoc group is more ambitious than reducing individual funds' carbon risk; they want to starve the entire fossil fuel sector of funds by convincing investors to exit the sector entirely. "When institutions imbued with public trust, such as Oxford and Stanford University, vote with their investment dollars to condemn the fossil fuel economy, it sends a message as powerful as any ballot box that the time has come to stop using the atmosphere as a free dumping ground."[47] The irony of the Rockefeller Foundation agreeing to divest of fossil fuel investments, which they pledged to do back in 2014, amplified the message.

On the surface, this appears naïve. Selling off Black Hat energy stocks at any volume that mattered would result in those stocks being undervalued and snapped up in an instant by another, less enlightened investor. However, not every investor need participate in a divestment movement for it to be effective. The cost of capital is dependent on the supply of that capital. If most, but not all, institutional investors shy away from Black Hatted energy companies, the cost of capital for those companies will rise, at least to some extent, given the inexorable effect of supply and demand. That's a signal impossible to ignore in such a capital-intensive business.

One outcome of carbon risk being more visible is to put pressure on energy majors to make a shift in the types of reserves they hold. Not all reserves are created equal: natural gas has half the carbon content of coal, and light oil up to a third less than heavy oil. The order in which fossil fuels will likely face terminal decline in demand are coal, then heavy oil, followed by lighter (traditional) oil, and last of all, natural gas. As companies like Equinor shift out of Alberta's heavy oil, others like Shell increase their commitment to natural gas. Exploration and investment focus can change for lots of reasons, but Shell and Equinor have been quite vocal as to why they're doing it: they're lowering their carbon risk. It's not as easy today as it was ten years ago to fund Alberta's heavy-oil patch or the coal sector. Investors are waking up to the long-term trend away from high-carbon fuels. That's a good thing.

What else might investors push energy companies to do? Stop spending money to find new reserves; use that capital for share buybacks. Whatever portion of existing reserves one might think we'll burn — one-quarter to stop at 2°C, maybe half to stop at 3°C — there's no way on God's green earth we can or will burn them all. Using up existing reserves will get us to 5°C–6°C, at which point all bets are off that the economy still functions at all. We can't burn existing reserves whatever the outcome. Hence, it's irrational to spend money to find new stuff since 100 percent of incremental reserves are unburnable. Currently, fossil fuel companies spend more hunting for those new unburnable reserves than they provide in dividends.

Fossil fuel stocks will eventually crash. Stock buybacks in lieu of exploring for unburnable carbon would soften that landing by reducing the total equity in play as the value of reserves declines in concert with increased carbon risk. It's hard for an energy company to step out of line; investors typically punish companies that don't work to replenish their reserves. But as investors better understand carbon risk, they'll do the opposite and reward companies that defund exploration budgets.

Increased transparency on carbon risk has little political cost, and it's highly robust — unlike a price on carbon, what's the motivation to undo it once it's in place? Its effectiveness in enabling investors to protect themselves from the carbon bubble is clear. Over the short-term, it may even force energy giants to switch to lower-carbon reserves, but the jury's out on whether it moves the broader investment community from fossil fuels into clean energy alternatives. Perhaps investors switch from coal to making T-shirts or building condos. Protecting capital from carbon risk is one thing, using it to solve the climate crisis is another. Saving investors' money as coal stocks crash is nice, but moving it to help the Cleantech Bulls is much more helpful. With so little downside, however, a federal policy to ensure transparency and standardization of carbon data is a no-brainer. Even the central bankers agree.

CLIMATE RISK: PLAYING DEFENSE

If the carbon bubble seems bad (and it is), it's nothing compared to the main event — climate risk itself. That is the big kahuna. As weather gets weirder — hotter, drier, wetter, windier, stormier — and as oceans begin to rise, untold economic seismic events kick in. Poor Puerto Rico saw it, as did wealthy Houston. The National Oceanic and Atmospheric Administration (NOAA) estimated 2017 economic losses due to climate in the U.S. alone at $306 billion. Less than half of that was insured. As these losses increase, private insurance (and reinsurance) will retreat, leaving the bill to corporate and public purses. Carbon risk is limited at the top end to the collective value of fossil fuel companies, but climate risk is limited at the top end to *everything*.

Playing defense against these more existential risks is possible. Just as the ability of climate science to attribute individual weather events to long-term climate trends has increased, so, too, has its ability to provide more granular data on changes that may occur at some given place. Climate science remains probabilistic (as does most risk assessment), but narrowing ranges of uncertainty about specific events is beginning to provide the kind of quantitative language most investors need in order to act. The move from qualitative warnings to quantitative risk assessment will happen slowly, but once it does, investors will react to protect their assets. They'll have no choice.

Some will disbelieve what the climate risk quants tell them, of course. They're free to do so. They could stay long on Miami real estate as flood risk increases; double down on beef production in the U.S. southwest as droughts become more frequent and severe; ignore increased political risk in countries where climate disruption pulls and stretches the social fabric. They may even get away with it for a while, passing the risk off to others also willing to go long on assets that presume a stable climate. But over time, they'll either lose or change their mind. As climate risk becomes more transparent, investors' actions will inevitably align with it. Investment culture will change as Bay Street and Wall Street shift from seeing climate as an irrelevant and fuzzy moral issue to one of immediate concern — *their* concern. Some will react fast enough to protect assets, others will not.

There are some pretty smart folks working to bring climate risk into the clear, actionable, and quantified language of the financial community. One example is the Climate Service, led by CEO James McMahon, which works directly with NOAA to start their analysis with some of the best climate models and most comprehensive data sets around. Using geographical grids of as little as seven square kilometers, they link probabilistic assessments of increased risk over some period of time — of fire, extreme heat, flooding, drought, high wind variance, etc. — to all kinds of other data sets and physical models to estimate direct costs to investors.

An easy case is flooding. Let's say a fund is looking to buy and upgrade a portfolio of real estate assets. The Climate Service can quantify the cost and probability, over a given range of time, of potential flooding. Start with readily accessible topological maps combined with precise descriptions of each building in the portfolio. Link those to predicted encroachment of coastal waters, increased rainfall, and flood plains. Then generate building-by-building estimates of damage, increased insurance costs, and flood protection requirements. Those total costs adjust downward the value of that real estate and the length of time an investor may be willing to hold some of the assets. All to protect investors from potential losses. But more importantly, these projections send a signal to the real estate community that the value of their investments is under *immediate* threat. The drop in value from climate risk occurs *today*, not at some future point when increased flooding actually occurs. Good risk analysis brings forward future potential events to today's investment decisions.

Another example might be the long-term value of a rail transport company. Not only are its tracks and terminals subject to a similar analysis, but increased heat stress on the tracks can be predicted with precision. Trains must slow down when tracks get hot. That brings a steep drop in productivity. Which again affects net *present* value. Other examples abound. The trick is to link predictive, fine-grained weather models to expected degradation of engineering or structural performance. A more complex problem is to address a large institutional investor's entire portfolio to quantify risk to long-term profits. That analysis must include the entire supply chain, and hence physical assets of companies all over the world. While daunting, given available climate data, it's just a massive database of these probabilistic risk curves applied to lots of physical infrastructure.

The Climate Service's grand ambition is to produce accurate and granular data that outperforms the existing standard-bearers for climate/ economic risk assessment, DICE and RICE. Built at Yale by Nobel Prize–winning economist William Nordhaus, the Dynamic Integrated Climate-Economy model (DICE) and the Regional Integrated Climate-Economy model (RICE) produce no actionable information for individual investors. Their output is limited to deeply understated, inaccurate,[xviii] and general warnings like "climate change has costs"; quantification is at the most macro level, expressed as a drop in GDP. Those warnings, if heeded, might have been useful twenty years ago. What's an individual investment professional supposed to do with that? It's no wonder fund managers have largely ignored this issue. They're flying blind.

By building a bottom-up model of highly localized financial risks linked to specific assets and expressed in dollars and cents, the Climate Service's data is *actionable*; investors can make defensible decisions today on the basis of it. In early stages of market development, the Climate Service intends to provide their more climate-savvy clients with some-thing they value most: climate "alpha," or data that allows them to outperform their peers. As this kind of risk assessment becomes stan-dard fare over the longer-term, it gives the entire investment community a way forward where now there exists only a fog of uncertainty.

Leading institutional investors are no dummies. True, it's been frus-trating to watch the financial community all but ignore climate risk thus

---

xviii One of the most startlingly inadequate aspects of these models is the way they treat economic damage from an unstable climate. The "damage function" takes the form Damage (as a function of Temperature) = $1/[1 + (T/x)2]$, where x is an almost entirely arbitrary "scaling factor" in degrees, calculated from a bunch of inputs to provide intuitively reasonable damage amounts for a very narrow range/low range of temperature increases. In DICE, x = 20.46°C. Not only does it have little rele-vance at levels of warming that we might worry about (over 1°C), but because it's so broad — damage across the entire economy — it might serve as a warning to policy makers but provides nothing actionable for investors.

far. But when those risks are put to them in a language they understand, and especially when they know *every other investor can see the same data*, their bets will get much smarter. Flood barriers will get funded to protect assets in place. Places that can't be defended, like Miami, will see their values plummet. The flippant climate skepticism and agnosticism I encounter so often when speaking to financial professionals will give way under the stress test of putting your money where your mouth is. Once climate agnostics and climate hawks can place competing bets in a context where the risks are clearly and transparently quantified, there will be a lot fewer agnostics and a lot more hawks.

But this is just playing defense — protecting investments from the coming storm. That's not enough. We need investors to play offense: deploy their capital to solve the problem itself. No mainstream investor, no matter how large or savvy, sees it as their job to "solve" climate. And frankly, no investor or institution is large enough to succeed even if they did. Solving climate requires that the entire investment community acts in concert, with a new foundation on which to base decisions. As we saw above, a price on carbon is the best conductor (although insufficient on its own). We might, however, be able to nudge some portion of that community forward in other ways.

FIDUCIARY OBLIGATION: PLAYING OFFENSE

From mining to manufacturing, forestry to pharma, aerospace to automotive, all companies need capital and answer to those who provide it. We might characterize the investment community in much the same way as we did the energy giants — as White, Gray, or Black Hats — depending on the type of investments they make, the support they provide to progressive companies in their portfolio, and the kind of activist stances they take as shareholders. Those who aggressively back carbon-intensive industries play the role of Black Hats. Those who move capital into large-scale clean alternatives are White Hats. But these two extremes currently represent a tiny minority of the sector.

Just like fossil fuel executives, the vast majority of investors make up the middle ground, the Gray Hats. There's a reason for that: professional investors take pride in not allowing political, environmental, or ideological factors to play a role in their decisions, aside from when they *directly* and *provably* affect investment returns. The prime directive for a money manager is to generate the highest risk-adjusted return possible, given their set of investment choices. Any other consideration is off-limits. Sticking your neck out to try to solve the climate problem, even if it's in everyone's long-term interest, just isn't on. Turning Gray Hats to White looks a bit hopeless from this perspective. It's just not in the DNA of financial investors.

There is a tiny gap in this picture, however. Like all investors, pension funds have a fiduciary[xix] obligation to generate returns, but over a very long time horizon. They need to generate those returns at known points in time (and in known amounts), often many decades out. Hence, they have a unique relationship to a healthy economy over many decades. It's easy to construct an argument that pension funds have a fiduciary obligation to play offense on climate. Unchecked climate risk entails ecological collapse over the coming century; absent a stable ecosystem, there's little possibility of a healthy economy decades out; without a healthy economy decades out, pension funds cannot meet obligations to which they're already committed; hence, they have a fiduciary obligation to make investment decisions in such a way as to reduce climate risk (not just avoid it). This may be a solution to Carney's "over the horizon" formulation of the problem.

It's odd to think of pension funds as having the longest time horizon of any of our institutions, including national governments that have been around for centuries, but in a practicable and actionable way they do. Governments play to the next election cycle. There are exceptions: the U.K.'s carbon commitments are enshrined in law that transcends individual election mandates, constitutions are meant to be (sort of) forever, etc. But as a pragmatic matter, it's very difficult to turn current decision cycles of governments to mid-century concerns. Yet those of pension funds already are — as a direct obligation to their members.

---

xix   To look after the best interests of their membership.

Another push on pension funds might result from direct communication with their beneficiaries. It's entirely possible that, if canvassed effectively, the thousands of teachers or public servants and municipal workers that make up our largest pension funds have a view on the kinds of long-term investments being made on their behalf. They may well advocate that, all else being equal, climate solutions be prioritized as an investment category. Of course, there is normally a firewall between investment managers and those whose money is being managed. And that's a good thing, as investment professionals are hired precisely to take the burden of decision-making off the hands of a fund's contributors. But maybe that's a mistake.

Obviously, one doesn't want investment decisions driven by a massive mob of well-meaning but financially unsavvy stakeholders! That's a recipe for disaster. But as an *additional* decision-making criterion, prioritizing climate solutions may not be a bad idea. Fund managers would obviously need to balance that expressed preference against a matrix of standard risk-mitigation and due diligence. But at least when investment decisions are made, the preference to find rational ways to play climate offense would be a voice in the room. And all else being equal, it might tip the balance in the climate's favor.

It's helpful to compare investment professionals with elected officials in representative versus direct democracies. In the latter, citizens are directly tasked with making decisions — a kind of eternal plebiscite. That's like managing your own money. In representative democracy, people elect officials to take on that decision-making role. At its best, we hand decision-making power to people who invest the time and effort required for considered judgment, backed by a seasoned public service. That's like a fund manager, backed by analysts. But elected representatives are not unanswerable to general concerns of the electorate. They're given periodic mandates to guide the high-level values that govern decisions. In the same way, individual beneficiaries can express their view on playing climate offense while leaving the details to experts. Blind spots[xx]

---

xx   The sub-prime mortgage crisis comes to mind, as does the visceral pushback against Roosevelt's New Deal during the Great Depression. See M. Heffernan's *Willful Blindness* for other examples.

result when professional classes spend too much time with like-minded peers. A periodic reminder of the larger context in which they operate can't be a bad thing!

Note the same issue exists, however, as with any problem that requires collective action: no single pension fund regardless of size or influence can possibly play the role of climate-fixer. How can one fund take on an oversized share of that burden? We can shift our argument in response: if the industry recognizes that its fiduciary obligations rely on maintaining a healthy economy past mid-century, then that responsibility translates to advocating as an industry for policies that allow them to act in concert to reduce climate risk — like a price on carbon. Pension funds, acting together, would surely be a powerful advocate for aggressive climate policy. I'm arguing they've a *fiduciary obligation* to do so. And if their beneficiaries were asked, they may agree.

Hence, we come back to the critical role of the public sector in mitigating the climate crisis. There will never be more than a few White Hats out there who act in our collective best interests even without policy signals, just as there are only a few early leading lights in civic society on any great moral issue. But laws and regulations ensure what was once rare and noble becomes common and normal. Regulatory measures from pricing carbon to clean fuel standards and building codes are the signals to which all investors respond. By working with far-sighted corporate leaders, governments set rules that define not only what kinds of activities are to be profitable, but also the very standards of good corporate behavior we later take for granted. Climate Capitalism will not, and cannot, rely on the voluntary and visionary actions of a few leading lights. It must be a broad and systemic economic response to the world's most pressing problem, coordinated by policy and regulatory signals that make what's right commonplace.

Corporate culture is changing, albeit slowly. Most business leaders today understand climate risk is real and anticipate some kind of regulatory response. The signal from boardrooms has changed from "not relevant to our business" to "we have to deal with this issue, find a way." Some firms are going so far as to tie executive compensation to internal emissions reduction targets. There's nothing like a bonus structure to focus one's mind.

And over just the past few years, I've noticed more people within all kinds of organizations — from real estate management firms to pension funds to oil and gas companies — empowered to find ways to move in the direction of the White Hats. For example, all else being equal, a reasonable business case for a clean energy or efficiency project gets done. That was not the case just five years ago, when the same project needed to be so compelling it was impossible to ignore. This is not unlike other corporate culture shifts, like fixing the gender imbalance — at some point, a new attitude gets normalized around the board table.

The climate debate is much more complicated than simply throwing rocks at corporate giants and blaming them for the mess we're in. The few obvious Black Hats aside, the reality on the ground for our corporate leaders is more complex. It's a mistake to paint them all with the same brush, and doing so does great disservice to many people who really are trying to solve a difficult and complex problem. Vilifying corporate America (or Canada, or Europe, or China) misses the point; generalizing and promoting anti-corporate sentiment is unhelpful. It does far more good to identify and support the White Hats, and nudge the Gray Hats to shift their direction. And vote for politicians who will give corporate leaders with a conscience the rules they, deep down, want to play by.

A carbon price, transparent data on carbon and climate risk, and an evolving corporate culture are all good things. But they're not enough. These kinds of unthreatening, incremental moves are inadequate this late in the game. How we might create a Climate Capitalism that short circuits an otherwise slow and stately pace to de-carbon is the subject of the next two chapters.

CHAPTER FIVE

MARKET INTERVENTIONS – ACCELERATING CLEAN TECHNOLOGIES

In our extended metaphor of the economy as a modern farm, carbon pricing is like the wires that guide our vineyard's growth. Transparent data about carbon and climate risk is akin to an annual pruning, keeping the plants focused on useful growth and open to air and sun. These initiatives make the plants healthier and more productive but are still not enough for our vines to produce as many grapes as we will need. We need nothing less than the equivalent of a new agricultural revolution. Similarly, carbon pricing and the other measures we talked about in the previous chapter are fundamental and useful tools in curbing climate disruption, but they are not nearly enough to get us to where we need to be in the time we need to get there. Climate Capitalism demands, and allows for, much more creative and aggressive ways to accelerate our path to a low-carbon economy. What follows is a miscellany of approaches to achieve that. All are interventionist, certainly, but I would argue they are a reasonable centrist's response, intended to keep our economy functioning effectively in the face of increasing climate risk — something everyone can agree is a good thing!

There are two kinds of market intervention: pulling existing solutions into the market (think fertilizer) and pushing new solutions into the market (think inventing better hybrid plants or farm equipment). In climate terms, the former is about meeting short-term emission targets using what's at hand. The latter increases the performance and lowers the costs of clean energy alternatives.

The sheer number of market gaps, or failures, identified below may imply a massive government bureaucracy to fill them all, but that need not be the case. For instance, there's a precedent for an effective nongovernmental institution — the kind of green bank I talk about — that brings the kind of discipline most commonly associated with the private sector. It offers a way to fill these gaps without the coarse blundering often associated with direct government intervention. There are also more efficient ways to fund closing the existing market gaps than direct public subsidies, a publicly backed but privately funded green bond, for example. While not free, it cuts the cost to taxpayers a lot. The green bank and green bond are pragmatic partnerships between the public and private sectors, leveraging the best each has to offer. As we'll see below.

A REGULATORY WORKOUT: FLEX!

There's no reason we can't address climate risk by regulation alone. One could, in theory, just ban emissions! But you can't regulate what can't reasonably be done. Such a hard regulatory approach would cause mayhem and economic collapse. A slightly softer, but still heavily regulatory approach is possible with foresight. If we knew precisely how technologies will evolve, we could ban all kinds of things in favor of their clean equivalents: EVs for traditional cars, solar plus storage for thermal coal plants, heat pumps for natural gas furnaces. But we are not so prescient and would, therefore, inevitably miss the mark. On the other hand, regulations can be so soft as to be merely aspirational: banning the sale of new diesel-powered vehicles, as proposed in France, for example. Or the voluntary national emissions targets of the Paris Agreement.

There is a middle ground. If a carbon price is champion of efficiency across an entire economy, flexible regulations (flex-regs) run a close second. More important, they're likely champions of effectiveness given a fixed political cost. Regulations generally come at a lower political cost than carbon pricing, since they're less visible to the voting public. If those regulations are also flexible, designed to be responsive to private sector innovation, they can be made more effective. Hence, there's a good argument flex-regs can maximize effectiveness while minimizing political cost, as we'll see in some examples below.

Don't confuse flex-regs with old-school command-and-control. This isn't the regulatory overreach of well-meaning bureaucrats. Flex-regs define an outcome but not how to achieve it. They're simple but legally binding rules imposed on a narrowly defined sector or industry. The degree of "flex" matches the kind of outcome you seek. At the highest level, a general emissions target for an entire country, for example, industries can best determine their lowest-cost compliance path. In this case, there is a large amount of flexibility afforded to how such a broad target is achieved. Applied more specifically, to existing technologies like heat pumps, for instance, emissions targets can be more stringent. There is inherently less flexibility in meeting this level of regulation because the parameters are so much more prescribed. But for stuff you want to target very narrowly, like methane emissions from fracking — which are twenty-five times worse for the climate than carbon dioxide — there's no need for flexibility at all. Just put a lid on them, no fuss no muss. The trick is to balance a regulation's degree of flexibility with its scale of ambition and ease of compliance.

Carbon pricing gets all the attention, but it's flex-regs that have been doing most of the heavy lifting so far, even in places with a carbon price, like California, Alberta, Quebec, and the U.K. There are two reasons they're a better bet than carbon pricing. First, they can be more effective, as they can be tuned to a specific industry's price and technical sensitivities. For example, demand for gasoline is price insensitive, so to make a dent, carbon pricing needs to be so high as to be politically impossible ($15 per ton on carbon adds only 3.5 cents per liter at the pump). Better to target fuels separately with low-carbon standards for auto fuel, for example. That short circuits price insensitivity and at the same time drags reluctant fuel producers into the climate fight since they're the ones that need to meet the standard. Second, flex-regs have a lower political cost.

Since flexible regulations specific to different classes of fuel are generally less visible than an economy-wide carbon price, they're not such an easy political target.

Flex-regs work best when solutions are already available. They're good at speeding up transitions that may otherwise take much longer: a building code to ensure that new buildings use heat pumps bypass the need for energy-savvy consumers; new auto standards for fuel efficiency diminish the effect of marketing monster trucks to urban dwellers; efficiency standards for windows, insulation, and light bulbs accelerate better products to market despite an industry motivated to keep selling old stuff.

Flex-regs are not always good at incentivizing innovation. It's unlikely the presence of a regulatory environment (however well-designed) will, in and of itself, conjure up new solutions that require capital-intensive production. For example, in the U.S., there has been a standing commitment to regulate up to 5 percent cellulosic ethanol in gasoline, backed by strong financial incentives. This seems like something that would spur new, clean fuel production, but it has not resulted in many next-generation commercial ethanol plants being built (a problem we'll address later in this chapter). However, flex-regs like these can be supportive of other policy initiatives that target early commercial, large-scale production, in that they guarantee a market for those who do manage to build those innovative clean fuel plants.

Even a regulation that appears stringent on the surface can be flexible underneath. Ontario phased out coal and Alberta is committed to eliminating it by 2030. That's pretty strict: no more coal! Ontario provided little option for market dynamics to play a role in its replacement; the Green Energy Act fixed prices for renewables, and the refurbishment by fiat of nuclear was equally rigid. But Alberta allowed market forces to define what mix of energy is best able to replace that coal. Natural gas is obvious, but Alberta also negotiated some of the lowest-cost utility-scale solar rates in the world. Over time, storage will play a larger role. The lesson is the more strictly defined the outcome (no coal), the more flexible should be the route to its satisfaction (replace it with what you will).

More than thirty states[xxi] in the U.S. operate under a renewables portfolio standard (RPS), which defines that some portion of their electrical supply must be provided by renewable energy. It's up to utilities to figure out how to reach that goal. They choose the mix of energy sources and the ways they contract for those energy sources. Canada's proposed Clean Fuel Standard (CFS) is similar. Fuels are differentiated by type — liquid, gas, or solid — and providers will be mandated to have some portion of their fuels be low-carbon. It's up to them what path to go — hydrogen, electricity, biofuels — and what technology they use.

Some regulations are a combination of aspirational and flexible. Typically, they define stringent targets many years out, combined with incentives to prod the market along in the near term. No less than nine European countries and dozens of cities are seeking to phase out internal combustion engines in passenger cars. No more engines — that's pretty strict! But it's decades out. Car makers, municipalities, and consumers can choose their solution — electric, hydrogen, transit, or biofuels — and the targets will likely get firmed up as solutions appear. The regulation itself isn't enough to force an industry to change, but the cumulative effect of more and more cities indicating a readiness to regulate, and consumers starting to make those choices, ensure that car makers pay attention now.

Flex-regs will continue to do the heavy lifting in mitigating climate disruption, partly because carbon pricing can only go so high, especially in the current political climate. The political right in the U.S. and Canada can play to their more radically anti-climate-action base for only so long. When it comes their turn to govern, they'll need to regain climate credibility to retain economic credibility. We're too far gone to ignore the long-term link between them now. Flex-regs give them an opportunity to do something that comes at a low political cost and a chance to crow about doing things more effectively. In Ontario, Doug Ford has already proposed increasing ethanol in gasoline from 10 to 15 percent.

---

xxi   Iowa was the first, and Hawaii is the most aggressive (requiring 100 percent renewable by 2045).

CAPEX OR OPEX? IT'S ALL IN THE HEAD – LESSONS FROM A LOW-CARBON HOTEL

From entrenched interests to engagement on climate risk in the investment community, we know attitude matters, even though in most cases it plays second fiddle to self-interest. But attitude can be a potent force. I'll start with a simple example.

Back in 2008, my friend Anthony Aarts asked me if I'd back his latest real estate project. He wanted to convert an abandoned building in downtown Toronto into Planet Traveler, a world-class hostel for travelers. The building was in a great spot, where cool Little Italy meets hipster Kensington. Anthony lives for lovingly restoring buildings and long had ambitions to build a peerless hostel. For him, it was a natural project. Those two interests meant little to me, so I gave him a challenge: if he'd agree to the goal of building North America's greenest hotel, I was on board.

By "green," I didn't mean recycled toilet paper or promoting the reuse of towels to save detergent — that's greenwash. It makes a hotel look good and customers feel good, but doesn't accomplish much compared with the scale of the problem. I was hunting much bigger game. I was aware back then that global carbon emissions had to come down by about three-quarters, so I decided to set that goal for Planet Traveler itself — a three-quarters cut in energy use — and see how far we'd get. You can't hit that target simply by reducing laundry loads.

To be honest, when we started, I had no idea how to hit that goal, or whether it was even possible. I like to joke that I didn't know geothermal from a hole in the ground (ha ha)! But as someone who builds companies for a living, I knew if it was going to fly or set any kind of lesson for others to follow, it had to be profitable. So that became part of the experiment; everything had to pay for itself. I would present options to Anthony for ways to reduce energy, and he would accept them only if they made economic sense. And that included not compromising the comfort of his guests.

I learned something quite profound, and that's the moral of this story: attitude really *does* matter. But first, I'll tell you how we did it. Turns out it's ridiculously simple. Right out of the gate, it's best to forget all the guidelines — all the green hotel codes and guides, and even the industry-leading environmental standard for commercial and industrial buildings, LEED. None come close to the three-quarters mark. They're really just a distraction. It turns out everything we need is already out there, fully de-risked and ready to go. We just needed to install it.

We started with geothermal heating and cooling — that's the workhorse of the hotel. Instead of using a furnace to create heat, you suck heat from the ground. And instead of air-conditioning to cool the building, you push heat back into the ground. Heat pumps move heat, instead of making heat. In our case, we use a series of pipes buried in the ground, which is super-efficient, since at about six feet deep, the ground stays about the same temperature year-round. (You can also move heat to and from the outside air using an air-to-air heat pump, but it's slightly less efficient.) While it sounds crazy, you get four times[xxii] as much energy out as you put in! This stuff has been around for decades — there's one in your refrigerator moving heat from the milk and butter into the kitchen! — and it works perfectly. Since that time, I've put heat pumps throughout my house in Toronto, which gets me off my natural gas furnace entirely.

There were a few other bits and pieces. We have some solar to preheat water feeding the hot water heaters, as well as some for electricity. The building is outfitted head-to-toe with highly efficient LED lighting. We can light up the entire building for less energy than we use in the four-slice toasters in the kitchen! Smart thermostats make sure we're heating only the parts of the building that are occupied. And finally, we grab heat coming out of the showers and going down the drain and recycle that heat to the showers that generate it in the first place. That's just a half-dozen simple, gravity-fed, passive copper-coiled pipes (called PowerPipes).

And that's it! That stuff is enough to hit the three-quarters mark[xxiii] in energy reduction. We didn't need fancy new inventions or trust in

---

xxii No, this doesn't break any laws of thermodynamics, because it's premised on moving (not creating) heat.

xxiii For the energy geeks out there, the relative contributions of the technologies are (roughly) LED: 10 percent, geothermal 40 percent, solar thermal 10 percent, solar PV 10 percent, and Powerpipes: 5 percent — for a grand total of 75 percent.

technologies that no one had used before. Everything was well understood, and already in use in thousands of other buildings. All we did was put it together in one place. Much to my surprise, hitting that aggressive target could not have been easier. And my penny-pinching partner was happy as heck with the result.

The only creative thing we did was face the problem of having nowhere to bury the geothermal pipes. The building takes up the entire footprint of the property, and there was no way to bring modern drill rigs[xxiv] into the basement. We engaged with the City of Toronto to let us use a public laneway which runs alongside the hotel. There was a bunch of permitting issues to get through, and it was the first time the city had encountered the idea of leasing their land for geothermal, but they eventually championed the idea. As a result of Planet Traveler, laneways and parks all over the City of Toronto are available for geothermal. That was the result of the leadership of Toronto's mayor at the time, David Miller.

Ah, I hear you say: easy enough if you can throw money at the problem! And it's true, there is a cost. But it turns out the total cost of these retrofits was less than 5 percent of the building's value (about a quarter of a million dollars). And all of them have a payback period — the time it takes for the energy savings to add up to cost of the equipment — that varies from six months to four years. Factoring in all the technologies, the payback period is somewhere between two and three years. Which means after that time, the systems are *actually making us money.*

Even better, instead of waiting three years to come out ahead, we borrowed that money against the building (some of the cheapest money around). The loan payments are less than the energy savings. That's the kicker: we were making more money from day one and are wealthier as hoteliers for making the decision to massively reduce energy. That's better than free. It's a no-brainer.

---

xxiv A company called Fenix Energy, in Vancouver, has since developed a fully self-contained drilling system that can be driven into the underground parking garages found in commercial buildings. It can drill down a mile in just over seven feet of headroom. So almost any commercial building in the world can now be retrofitted with super-efficient geothermal heat pumps.

I developed a rule of thumb from that project: leveraging just 5 percent of a building's value to pay for energy-saving retrofits would lower the energy use in a building by about three-quarters and pay for itself from day one if the money was borrowed against the building. That rule of thumb appears to scale up to the largest buildings. When Honeywell revamped the Empire State Building in NYC, they spent the equivalent of just under 2 percent of the building's value and lowered energy by a third — and they accomplished that without the heavy lifting of geothermal.

But this story is not really about that little hotel. There are two larger morals to the story.

First, this project proved just how easy it is for us to reduce emissions on a big scale. Buildings account for almost half of greenhouse gas emissions. If everyone did what we did with Planet Traveler — and they can — we'd lower global carbon emissions by a third. That's more than enough to meet our Paris commitments. And everyone who did it would be richer! And almost all of that economic activity remains in the local economy: local trades, engineers, and material form a bulk of the expense. This fruit is so low hanging all we have to do is bite. The only barrier is attitude. I was a highly motivated building owner and drove the engineers to find energy solutions. And I was willing to borrow against the building to get it done. That kind of mindset and purpose is available to anyone.

While we were building our hotel, the financial crisis of 2008 hit with full force. Governments across the globe were working to find ways to inject capital into the financial system and stimulate economic activity. Our little hotel was an example of how to get that stimulus right. Instead of throwing cheap capital at massive institutions loath to increase lending in a down market, public funds to retrofit building stock would ensure stimulus funds flowed to small business and local employment.

Which leads to the second, larger lesson, one that should be taught in our business schools. As a hotelier, I saw energy costs as strategically important. I looked upon that line item on our income statement as one of long-term importance, not as an afterthought. That's why I was willing to borrow against the building — to tap our capital budget — to get it done. Had I viewed energy costs as just another expense, we'd have had to find the money from our operating budget. That typically means only projects that pay back *within a year* would make the cut because you have to find the cash from operations.

CFOs are taught that only those expenses related to their core business are eligible for access to the cherished long-term capital budget. If you're a T-shirt company, it's new sewing machines. If you make cars, it's machinery to produce car parts and/or put them together. Very few see energy costs that way. Which means, absent a highly motivated executive willing to push the bounds of normal business behavior, energy efficiency requests are part of the operating budget. Which means only those with a payback of a year or less get done. That's a tiny fraction of what we could be doing — the equivalent of limiting the hotel retrofit to LED lighting and smart thermostats and nixing the rest.

All that's required to light up the energy efficiency sector is for CFOs to see long-term energy costs (and carbon risk) as being of long-term strategic importance. It blows my mind that business schools have largely missed this critical piece of education. This unlocks the capital budget, which lengthens the payback time, which in turn enables much deeper cuts in energy use (and cost savings). It really is as easy as it sounds.

Here's the craziest part: energy efficiency projects generally make better returns than the core business itself! Any company generating more than 10 percent on its capital is doing great (8 percent is typical), yet energy retrofits, even those with longer payback periods, easily double that number.

That hotel represents just one flavor of energy efficiency. Others exist across the economy — in our factories, resource sector, transportation fleets, cement plants, and mines. Better analytics open up lots of ways to extract those efficiencies: throw a bunch of sensors on operating equipment, and let machine learning teach plant operators how to minimize energy.

There is no credible low-emissions scenario — from the World Bank, International Energy Agency, IPCC, anyone — that doesn't assume energy efficiency does at least half the heavy lifting of climate mitigation. Unlocking that action often requires nothing more than a change in attitude in the community of CFOs: long-term energy cost and carbon risk are of long-term strategic importance. And companies that deploy capital to make operations efficient don't just get more profit. They become resilient to carbon and energy risk.

FIRST ON THE FLOOR: PUBLIC SUPPORT OF THE BIG STUFF

Imagine a high school dance. The first few songs come on the loudspeakers. Every kid really wants to get out there and dance, but no one wants to go first. The brave few who get out on the dance floor take the risk of being embarrassed and looking silly. Everyone knows you can just wait for others to lead the way. By following them, you get all the benefits of dancing the night away with none of the risk of being singled out. The first ones on the floor take all the risk, but the benefits are shared by everyone. Luckily for all of us, there are always a few kids willing to leap about or shuffle quietly on their own. That dynamic is not so different from certain kinds of technology development: lots of companies may benefit from it getting done, but no one wants to take the risk of going first.

The private sector is best motivated to spend money on technology development when they're sure to reap the rewards for taking on the risk and cost. The most obvious way to protect that investment is with intellectual property — filing a patent. But patents don't always apply. Some kinds of technology, particularly those related to large energy projects like hydro, are better thought of as a set of engineering competencies. Others involve such long-term risky research that the rewards are too far out to matter. Think nuclear power last century, where all the basic designs were inherited from the military establishment. When patents don't apply, or the risks are too high and rewards too far away, the private sector tends to be absent. Hence the public sector is critical. Like those first few dancers.

While I think distributed energy plus storage is the long-term winner (that's where ArcTern places its bets), it would be prudent to ensure large-scale low-carbon baseload[xxv] sources are in the race early on. The "big

---

xxv Baseload power is twenty-four-hour-a-day, seven-days-a-week power typically associated with large electrical plants, like those that burn coal or nuclear fuel.

three" contenders — carbon capture and storage (CCS), next-generation nuclear[xxvi] (NGN), and enhanced geothermal systems[xxvii] (EGS) — are just that: clean sources of industrial-scale energy. None of these is going to be developed by the private sector acting alone. Either the risks are too high, the rewards not protected, or the time frame too long. Which cries out for public sector action.

The history of CCS makes a great example. CCS provides a lifeline to existing energy companies since it can, in theory, make coal and gas plants carbon-neutral. Given the massive balance sheets and depth of engineering expertise in the fossil fuel sector, you'd be forgiven for thinking these companies would be motivated to bring this technology to market, and capable of doing it. But very few companies have invested in it absent massive public funds. The few CCS projects that have been developed relied on huge injections of public money. This isn't just industry holding out, knowing they can get free cash. Nor is it because early deployments are uneconomic, although they are. There is a deeper disincentive.

---

xxvi By NGN, I don't mean fusion power, although the same argument applies. Instead, there are existing fission-based designs sitting deep within the network of U.S. national labs (particularly Sandia National Laboratories), shelved decades ago in favor of current commercial designs (chosen to provide the military with weapon-grade material). NGN are modular, passive-safety breeder reactors being commercialized by the likes of Canada's Terrestrial Energy, which aims to commercialize a molten-salt reactor. NGN reactors recycle nuclear waste, turning it from a problem to an energy source. They "burn" it multiple times to extract an order of magnitude more energy than we got the first time. The final waste has a half-life measured in hundreds of years, not hundreds of thousands. Their design addresses today's safety concerns: there's no active cooling, hence there's no danger of meltdown.

xxvii EGS taps the high heat found deep beneath our feet, almost anywhere, to make electricity. It means drilling deep down (six to ten kilometers) to tap into hot, dry rock. The rock is fractured, much like we do now to release shale oil gas, but here we go deeper and extract heat instead of gas. The fractured rock creates a loop between the two drilled holes, through which water is pumped. It turns to steam, and that energetic heat drives a turbine. Experimental plants in Europe and Australia have proven it works. The energy potential is astounding: available to us is somewhere between three and thirty *thousand* times our energy needs as baseload power.

Like large-scale hydro, CCS and EGS are a set of engineering techniques and competencies, rather than patentable discrete technologies. These energy sources are unlocked not by inventing a new gadget, but by developing a set of skills and knowledge. They are experiential knowledge more than protectable intellectual property. Crucially, any company that goes through the pain of acquiring the requisite knowledge will inevitably share it with their peers. They take the pain but share the gain. Pioneers are disadvantaged because competitors benefit from their risk. And, therefore, few are willing to blaze that trail on their own dime.

There's lots of public funding for large-scale demonstrations of CCS in Canada and abroad (although the U.S. recently backed off). The powerful fossil fuel industry is able to extract public support for this initiative, which works to their obvious economic advantage. However, for an emerging industry or technology, that is less the case. So EGS remains orphaned, despite the enormous potential upside.

One way of breaking the impasse is to copy ways in which industry has found to share experiential knowledge. In Alberta, most heavy-oil producers face the same two problems: the carbon content of their oil is the highest in the world, and many produce lots of toxic effluent.[xxviii] As an industry, they stand or fall together on these risks. They're motivated to solve them together in a coordinated way. As a result, they formed Canada's Oil Sands Innovation Alliance (COSIA), a vehicle through which they share technology, techniques, and intellectual property. Each player contributes to a common cause.

A similar structure would work to get EGS off the ground. The challenge is the absence of a large, preexisting industry from which an alliance might be formed. The public sector can and should take the lead on this, providing capital and a forum in which partners collaborate. They're the lead dancer to get others on the floor. In 2011, a group of brainiacs at the Waterloo Global Science Initiative's (WGSI) Equinox

---

xxviii Open-pit mines produce the effluent that makes up huge tailings ponds (more like lakes, really), not the in situ operators, but it can be argued they all suffer from the public relations nightmare that comes from the open-pit operators.

Summit produced the Equinox Blueprint: Energy 2030. In that report lies a key recommendation yet to be acted on.

WGSI advocated a public-private consortium funded with enough capital to develop ten small commercial-scale EGS plants around the world. Share the resulting knowledge and expertise: fracking methods, ways to maximize energy throughput and efficiency, and a mapping of the resource. Partners are those who would most benefit from, and be able to act upon, reduced deployment risk. Each plant costs about $100 million, so total funding is about $1 billion — a drop in the bucket compared to the public support given to CCS. Only when those EGS plants operate successfully will there be enough certainty about cost and reliability to open the taps of private capital.

Next-generation nuclear could use something similar, although it's less about sharing data and engineering techniques than providing financial and regulatory support: money to build commercial-scale plants and a fast-tracked regulatory environment that allows for unique one-time builds that demonstrate technical and economic viability. Tomorrow's nuclear plants need not be today's expensive monoliths. And with standardized designs, a different safety profile, and a ready source of fuel, they're well positioned to play some role in our climate response. With the added benefit of using nuclear waste, they're not a bad bet. Public fear of nuclear stopped development on breeder reactors[xxix] back in the 1970s and continues to stymie progress today. That's a mistake.

My own view is that none of these will compete against a distributed grid powered by renewables and storage: CCS is dead on cost alone. NGN takes too long. Of the lot, EGS is my favorite. But that doesn't mean we shouldn't accelerate *all* of them into market. Most obviously, I could be wrong. More important, when the planet starts to get hotter, we'll want as many potential aces stuffed up our sleeve as possible. A dozen operating plants of NGN, CCS, and EGS extends our options. Only the public sector can get that done — just as it did for our existing nuclear fleet.

---

xxix A breeder reactor can take spent nuclear fuel and extract more energy from it, iteratively, to result in much more power from the original fuel and much less long-lasting waste.

NEGATIVE EMISSIONS: TURNING BACK TIME?

For centuries, we've been digging up fossil fuels and expelling the embedded carbon into the air. Obviously, we need to stop doing that. Unfortunately, it's equally obvious that gradual reductions can no longer be enough to make up for historical emissions. We'll likely overshoot our shared goal of 2°C by some margin. What to do? In theory, we could reverse that process and effectively go back in "emissions-time." Instead of extracting and emitting carbon, we suck it out of the air and stick it back in the ground. Sounds far-fetched, but most of the livable[xxx] IPCC scenarios assume we do just that.

Carbon capture suffers similar criticism to geoengineering: it can be used as an excuse for not doing the real work of reducing emissions. If carbon capture offsets emissions that could have been avoided, then it makes no mathematical (or economic) sense. One person pollutes, another cleans it up. And the cleanup is far more expensive than avoiding the mess in the first place. Imagine one person flinging trash as they drive along the highway. Another is tasked with cleaning up all that garbage. It's surely cheaper for the litterbug to buy a garbage bag than it is for the rest of us to hire the cleanup crew. But when things get hot and we panic, we'll need *all* options on the table, including EGS, NGN, and CCS.

There are two ways to suck carbon out of the air: natural and synthetic capture. Trees and plants naturally take up carbon to use in photosynthesis. Afforestation on a huge scale, billions of trees, would make a difference. Tim Flannery[48] has proposed massive planting of fast-growing switchgrass; each season it sequesters carbon in the ground via annual root growth, roughly equal to the height of the grass. Switchgrass planted on marginal land has more potential than reforestation since more unproductive land is suitable and the carbon uptake to soil is faster. Oceans also take up CO_2 at a natural rate, which we

---

xxx   The most aggressive scenario, which is purportedly safe, relies on massive emission reductions *and* capturing atmospheric carbon later in the century to compensate for overshooting those safe levels.

might increase by seeding them with algal accelerants like iron filings. The first two options above suffer from a scarcity of land, the latter from all kinds of ecological complications.[xxxi] In the absence of carbon pricing, all three suffer from zero market incentives.

Synthetic processes take two steps: capture carbon dioxide by some chemical process (the active agent is called an absorbent), followed by releasing the captured gas to a permanent container or combining it with something to make synthetic fuels or other materials. Some processes capture low-concentration CO_2 in ambient air (direct air capture), others capture it from high concentrations found in industrial exhaust stacks (stack capture). Direct air capture has the disadvantage of having to move massive amounts of air since CO_2, although effective as a greenhouse gas, makes up a mere 0.04 percent of outside air. Eleven Astrodomes hold one ton of CO_2. If you processed air filling the entire Grand Canyon, you'd get less than 130 tons. A hybrid natural/synthetic approach uses fast-growing biomass, like switchgrass, as fuel for electricity production combined with capture and sequestration (BECCS, biomass energy with carbon capture and sequestration). All of the optimistic Shell scenarios rely heavily on BECCS.

These proposals differ from traditional CCS because they permanently store carbon captured *without* burning fossil fuels. That's what turns the clock back — they're carbon *negative*. The problem is only a sufficiently high carbon price can create this market. In the meantime, the Carbon XPRIZE motivates innovators with $20 million in cash for those who can make commercial (if niche) products from captured stack gas. Innovations range from injecting CO_2 into cement blocks to make them harder, feeding it to algae to make fish food and nutraceuticals, and even using it to make carbon-neutral vodka. While my heart's with the vodka, the smart bet is on cement injection.

---

xxxi   Algal blooms, which certainly sequester carbon, will follow if iron filings are dumped into the ocean, as a rogue inventor confirmed off the BC coast some years ago. Iron filings have a short resident time in shallow water, however. Even if one overcomes that via binding to a lighter medium, there is no guarantee the resultant algal bloom will descend to depth — a requirement if the carbon is to remain sequestered.

There are two independently funded contenders way out in front of the Carbon X crew. Europe's Climeworks uses a solid absorbent to make modular, small-scale, carbon-sucking machines. They started with ten machines that feed CO_2 into greenhouses to increase photosynthetic production.[xxxii] Canada's Carbon Engineering (originally backed by Bill Gates and more recently by Chevron and Occidental) uses a liquid absorbent as a first step for direct fuel production (air to fuels).

The scale required to make a dent in emissions-time is staggering. Climeworks's laudable goal is to grab 1 percent of global carbon emissions within a decade. That means building five million of those machines that feed greenhouses. They've built 100 so far. And they need to find a way to permanently sequester the carbon instead of making slightly bigger tomatoes. Carbon Engineering faces similar scale challenges. This shouldn't be surprising. The industry they're comparing themselves to is the largest in the world, built over a century. Is it possible? Sure, in theory. The automobile industry makes tens of millions of cars a year. Why not similar numbers of these machines?

Estimates of costs for these technologies are hard to confirm since they're private companies. Both claim their economics work with a carbon price of $100 per ton and another decade or two of performance improvements. It's possible carbon pricing gets there in that time frame. And technology cost curves (like solar) often surprise even the deepest skeptic. But those cost reductions depend on big increases in demand that bring production scale and ongoing investment in process and technology. Moore's Law doesn't operate independent of hard-driving market demand. There's a chicken-and-egg problem to crack here.

Increased demand can *only* come from a high carbon price (or some equally expensive regulation). Carbon capture is the ultimate environmental service; the product is limited to a broadly shared public good.

---

xxxii Climate skeptics continually repeat the trope that an increase in CO_2 is a net benefit to mankind since plants become moderately more productive. This myth has been repeatedly debunked. While there may be marginal benefits within a few agricultural niches, there's no evidence those benefits outweigh losses from increased climate extremes.

Hence, the creation of their market is entirely a political issue. There's no free market that could enable the negative emissions industry (current systems that generate value by using CO_2 to squeeze more out of old oil are not carbon-negative). Any skepticism toward carbon capture is not technical, but political. Equally, optimism depends as much on a belief in political will as the smarts of these entrepreneurs.

Let's take a conservative view and assume all carbon capture technologies require a floor carbon price of several hundred dollars. That's well past the point where almost all fossil fuel emissions are curtailed by clean alternatives; hence, it's not economically efficient for carbon capture to offset coal or natural gas plants, because one can assume by that point something else would have presented itself as a cheaper solution to replace those energy sources. So why might we need it? What's the point of backing carbon-sucking technologies that are so expensive to operate?

First, even if we manage to eliminate most fossil-based energy sources, we'd still need carbon capture. International jets aren't going electric anytime soon, so kerosene will be around for a while. And it's tough to replace emissions produced as an integral part of an industrial process. To make steel, iron, cement, and glass, we still rely on chemical reactions that emit CO_2. Even the Haber-Bosch process by which we make fertilizers releases emissions (that industry by itself adds up to nearly 3 percent of global emissions). So while carbon capture seems far-fetched, far off, and expensive — we still need it.

Imagine a cement producer in Canada paying $100 for each ton they emit to a solar-powered capture plant in Morocco to bury a ton there. Or me paying a carbon capture plant[xxxiii] in India to do the same for the emissions required to fly from Toronto to London (personally, I'd be happy to pay a 10 or 20 percent premium for the peace of mind that comes from moral clarity). The first deal can't happen without a robust global carbon market. Not an intrusive world government as feared by the neocons, just good carbon accounting. The second might be conjured

---

xxxiii It doesn't really matter if I pay extra to have the plane I'm on run on the kind of synthetic jet fuel produced by carbon engineering, or on regular kerosene and I pay to capture and bury the emissions. It amounts to the same thing.

up by private actors who link customers and carbon capture projects, but oversight is needed for anyone to trust the system.

Neither of those examples puts us back in time since they're not carbon-*negative* but carbon-*neutral*. They only slow the clock. And don't forget, all those carbon capture plants need to be powered with zero-carbon energy to net out at carbon-neutral. It's hard to see an economic argument outside the following: if carbon levels look to be crossing a super-critical threshold, we may be willing to pay astronomic sums to pull that number down. Recent evidence suggests, for example, that past 1200 ppm atmospheric carbon, the clouds disappear. Which immediately shoots us up another 8°C. One cannot put a limit on what we'd pay for insurance against that very thin, bright red line.

That's the ultimate use of carbon capture: not a magic bullet but an expensive emergency backup. Independent of its role as disaster insurance, it does poorly in each of our four policy characteristics: effectiveness, efficiency, political robustness, and cost. Given such a low score across that spectrum, policy to fast-track carbon capture is best seen as insurance, that's it.

So, what conclusions can we draw from all these examples of alternative technologies? Although carbon pricing can motivate this market, it's not enough — and it's instructive to explore why. For *any* cleantech innovation to reach climate-relevant industrial scale in a decade or two — whether it's carbon capture, energy storage, next-generation biofuels, or novel solar materials — it needs an extra leg up. Particular market interventions are needed to accelerate disruptive technologies that target industrial-scale infrastructure. That's because innovation and infrastructure simply do not mix — for many reasons. We can learn from the success of industrial-scale solar.

FREEWAYS AND ON-RAMPS

Solar energy has come a long way since the first basic solar panels were invented in the 1970s. Back then, solar was expensive and represented a tiny portion of energy installations. Some users were driven by a hippie feel-good vibe with a moral urge to get involved in a cleaner form of energy. Others were driven by a sense of independence, either because they needed to power a house in the middle of nowhere, far from the electrical grid, or because they felt the need to counter what was felt to be a growing energy crisis driven by an increasingly assertive and angry OPEC. Either way, solar was a small niche market.

Today's solar is a massive industry, with hundreds of billions of dollars of installations annually across the globe. It's now the single largest source of energy being built, outpacing nuclear, natural gas, oil, and coal. Factories in China pump out a mind-numbing number of panels — the equivalent of a thousand football fields *every day* — which eventually find their way to rooftops and solar farms around the world. It's often the lowest-cost new energy source, hitting record lows year after year as production costs drop and banks step up with super-low-cost financing[xxxiv] for utility-scale solar farms. If solar was once a back-country road, it's now a superhighway.

Things can only get better, right? Wrong. The success of traditional solar has the paradoxical effect of blocking new technology from getting to market, which in turn puts a limit on how much solar can keep improving. Why? Those giant factories in China may be modern manufacturing centers, but they're still producing the same old panels invented back in the '70s. To compete, a new technology has to get to an equivalent scale — both in manufacturing capacity *and* in the amount and cost of bank financing for customers. But incumbents are so big, investors are loath to build anything that tries to compete. And without a factory already operating at scale, no bank would consider low-cost loans to customers. It's like a multi-link catch-22: no factory, no debt financing; no debt financing, no customers; no customers, no investors; no investors, no factory.

Think of today's successful solar industry as a four-lane highway where cars are zipping by at a top speed of sixty miles an hour. All those

---

xxxiv The cost of debt on a solar project has almost as much effect on the final cost of energy as does the cost of the panels themselves. It wasn't until low-cost panels (from factories operating at huge scale), combined with banks' willingness to provide large amounts of low-cost debt, that the cost of solar began to come down to where it is today.

cars are models designed in the '70s. Promising new solar technology is like a modern car with the capacity to go a hundred miles an hour. But it's stuck trying to merge with the glut of older vehicles, waiting in vain for a break in traffic. The highway is filled with out-of-date vehicles limited in their top speed by their vintage, but collectively going at a high-enough speed to block new and improved cars from entering traffic. Just like the new fast car that can't get on the highway, new solar tech is stopped cold, stuck in the lab. And the entire industry is stuck in the '70s thanks to its own success.

The same can be said of other kinds of cleantech, like batteries. Lithium-ion (li-ion) battery systems, like those produced by Tesla's Gigafactory, are backed by global suppliers of the individual cells (like Panasonic) and dominate the energy storage market. Similar to the Chinese manufacturers' role in solar, Tesla brought down costs significantly. Which is a wonderful thing, but there are limits to where li-ion batteries can take us — on cost, certainly, but more importantly in terms of performance: energy density, cycle life, temperature sensitivity, and overall efficiency.

Energy storage markets will benefit from strong competition to li-ion. Yet investors look at the scale of Tesla (and other li-ion battery producers) and are gun-shy about supporting the development of new battery chemistries. And even if venture firms can bring competing chemistries to the point where they're market-ready, they still face the same two hurdles to commercial scale: the company needs to finance big manufacturing and find bankers to provide long-term debt to support their customers' deployments. The markets will not do this on their own. A lot of promising technologies will wither on the vine. Sadly, some have already done so.

In another era, this may not matter. So what if new technologies struggle to scale up quickly? That's life in a market economy. But the urgency of climate risk requires we accelerate new clean technologies into the market. Not only do we need continued drops in cost but also improvements in performance that can come only from new and better materials, radically disruptive industrial processes, and so on. We have climate solutions today, but they cannot take us all the way to zero carbon. We need a strong bench of upcoming innovation to constantly squeeze out incrementally difficult slices of our emissions pie. And we need those innovations in years, not decades.

We need to build on-ramps to the cleantech highway. We must facilitate the entry of competing technologies into fast-moving, high-volume markets that require a combination of capital to scale production and debt to finance deployment. To design a good on-ramp, let's look to how solar got to be a superhighway in the first place.

Back in the 1950s, solar cells were used to power American and Russian satellites. The basic technology was invented in a mix of private and public labs. Super expensive, but worth the cost up there in the heavens. In the 1970s, researchers found ways to lower the cost from hundreds of dollars per watt to about twenty. This effort was driven (somewhat ironically) by Exxon, who wanted to power lights on oil-drilling rigs out at sea. By the 1980s, solar energy expanded to include hippie and off-in-the-woods markets. It was put on roofs either to feel good about oneself or because there was no other way to power your lights and radio. It was expensive, had no real muscle, but it worked. Then Europe got in the game, especially Spain and Germany, who subsidized much larger deployments with feed-in tariffs[xxxv] in an early climate effort. That demand dropped the price by half. That gave solar some muscle, but it was still subsidy-driven.

Then along came China, who not only increased demand with massive solar farms, but — crucially — changed the economics of production by providing lots of cheap (even free) capital to solar panel manufacturers. The combination of large and growing demand by solar farms, combined with the influx of capital for supply, caused prices to plummet more than 80 percent in just five years. Today, China dominates panel production. Most of their factories don't make any money, the margins have been driven to near-zero. Many will never pay back the state-subsidized loans they were built upon. But for China, there are two strategic benefits. The first was to dominate solar, which they do. Second, they keep the factories

---

xxxv A feed-in tariff (FIT) is a guaranteed price for energy, backed by either an increased cost to consumers through the utility or, more often, the public purse.

humming at low or zero margin because they value social stability. That means keeping people employed.

The solar superhighway runs on technology invented fifty years ago.[xxxvi] It was built by a combination of European policy that initially increased demand, followed by a massive influx of Chinese state-backed manufacturing capacity. The result is a jaw-dropping crash in costs of almost three orders of magnitude in less than twenty years. Solar is but one among many examples of incumbent technologies blocking competition through sheer scale. Canada would do well to think carefully about what economic niches we seek to occupy in subsequent seismic shifts. Do we compete head-on with Chinese manufacturing or find a way to feed Canadian innovation into that industry? Do we allow foreign corporates to pick off promising Canadian biofuels technology, or leverage our geography to give that industry a head start at home?

As we think about how to position Canada's burgeoning cleantech industry for global success, we start by understanding the dynamics of existing cleantech superhighways like solar. For new technologies to have a chance of competing against incumbent juggernauts, we must find ways to get riskier but higher-performing technologies up to speed. That means building on-ramps unique to the characteristics of targeted subsectors. But instead of having governments *pick* the next technology, it's better for them to put in place structural supports to *back* companies that show they're capable of winning a market. Backing winners is not the same as picking winners.

OWNING THE PODIUM: BACKING OUR WINNERS

When Vancouver hosted the 2010 Winter Olympic Games, Canada had a record medal haul. It wasn't an accident. A few years earlier, the federal government started an initiative called Own the Podium. A select committee identified internationally competitive athletes early on — those

---

xxxvi There have been incremental improvements, of course, and some new materials coming into the market. But the dominant technology remains mono- and poly-crystalline silicon — old-fashioned stuff.

who had proven themselves in international-level competitions. They were provided enough funding to give them some independence from their normal working lives; they could quit their day jobs to focus on training. We also made sure they had access to great coaching. The result was that record medal haul. The Canadian government didn't pick the potential winners — it backed them.

We can take a similar approach to cleantech. The market is massive and growing. Countries, such as South Korea, Germany, China, and Japan, are aggressively supporting their cleantech startups with the goal of long-term economic upside. They are being very strategic. Germany, for example, often provides free engineering services as part of oversight of their development funding. Naturally, many of the systems just happen to have German-made components. Canada needs to do something similar and do it smart. Cleantech is a race and the stakes are high. If Canada gets just our pro rata share — 1.6 percent — of the global cleantech market, then by 2030, that sector would be larger than automotive. And pro rata is what you might expect just for showing up. Surely, we can do better than that!

If we can identify companies capable of delivering economically competitive solutions in large global markets, and back those companies with highly targeted support, they stand to gain an outsized share of their market. We'd do well to accelerate our cleantech exports if we want the equivalent of a record medal haul. This isn't about government picking winners, but *backing winners*.

The Americans got something right under Bush and Obama. The U.S. Department of Energy was tasked with accelerating American clean technology into the market. The signature piece of their response — called Title 17 Innovative Energy Loan Guarantee Program — was designed to provide exactly the type of on-ramp discussed above. By guaranteeing private capital deployed to accelerate new clean energy companies and technologies, it removed exactly the barrier to scale felt most acutely: that of deploying project-level capital for something new and relatively unproven. Their stated mandate was to "encourage early commercial use of new or significantly improved technologies in energy projects," and they did that by providing guarantees for loans and financing from the Federal Financing Bank.

The GOP made hay out of one of the more public failures — the infamous Solyndra debacle — but what they (and their teeth-gnashing punditry partners) failed to note was the overall success of the program. That program helped launch some of the most successful companies in America today, like Tesla and SolarCity. It also enabled the first five large solar projects over 100 milliwatts, which seeded the market and gave American project developers the heft and experience they needed to go global. In both cases, private markets have taken over, and the Department of Energy (DoE) backed off. The program directly deployed about $16 billion, but levered up more than $50 billion. The cost to taxpayers is puny, compared to the economic activity it generated and the long-term advantage some of those players (like Tesla) have in the market: the overall program suffered losses of only 3 percent, since the vast majority of loans were repaid.

Canada made a move in 2017. About $1.4 billion of federal money was put into play to accelerate our own champions. Most of it went through crown corporations Export Development Canada (EDC) and Business Development Bank of Canada (BDC). EDC got $450 million to support first-time deployments of high capital expenditures (CapEx) cleantech. The poster child during prebudget discussions was a next-generation biofuels plant. The jury's out on their ability to execute on the mandate. EDC is hamstrung by an agreement with other export development banks: no subsidies. But the task here is to move the market, to make something happen — the building of a first-time high CapEx plant — that would not have occurred in the ordinary course of business. Why else put the policy in place if not to change market dynamics? That requires, by definition, a subsidy — regardless of trying to call it something else. It seems to me the feds got the idea right, but the execution wrong. EDC's inability to provide subsidies is in direct conflict with the political mandate.

There are creative solutions. The most obvious is for EDC to act first and apologize later. Canada's not so good at that. Another is to ask for long-term upside (e.g., financing rights on a dozen commercial plants at above-market rates) in return for taking increased risk the private sector avoided. That takes creative thinking and motivation within EDC. We'll see. A terrible outcome would be to redefine cleantech and push the capital at traditional fossil fuel companies using it to clean up internal operations. That's not the type of zero-carbon technology the funds were intended for.

The danger of fossil fuel companies hijacking well-intended policy like that which empowered EDC is very real. Incumbents are good at capturing programs aimed at upstart competitors. One way around "program capture" is with a better definition of cleantech — perhaps "a greater than incremental improvement over incumbent infrastructure in decoupling economic activity from environmental harm." Or better, ensure the capital gets delivered to companies without access to a large balance sheet; put a cap on company size. That keeps the capital for newcomers, not incumbents.

More federal support for cleantech is expected in 2019, by which time this book will be published. One way forward is to find more creative ways to leverage the public purse to unlock the big chunks of capital waiting on the sidelines, including both development banks and the private sector. Many of my peers in the cleantech sector were convened for the CleanTech Economic Strategy Table (CTEST) to provide recommendations on how to structure that support. An Own the Podium–style model is a critical piece of our report, with a focus on support similar to what the U.S. DoE offered: "Public scale-up funding should be used to significantly de-risk clean technology deployments and drive down the cost of capital by focusing on large projects and on non-equity-based financing for later-stage companies . . . that are not yet bankable as defined by market lenders."[49] Bang on. Let's hope the implementation is better focused this time around.

Just as the original Own the Podium provided more than just money to athletes — such as a better coaching system — so, too, these recommendations target more than just money. The Trade Commission can be better leveraged to identify and enter new markets; we might aggressively support Canadian participation in the development of international standards involving cleantech; and we can try to better identify global clean technology talent that firms can draw on to support exports.

There's a good reason the Canadian focus is on exports, exports, exports. It speaks to how our economic self-interest and environmental aspirations are directly connected. Climate do-nothings complain Canada makes up less than 2 percent of global emissions, so why bother? The flipside of that argument can only be to aggressively affect the other 98 percent! There's no better way to do that than to export innovation solutions that out-compete fossil fuels: better energy storage, cheaper solar, cellulosic ethanol that beats gasoline, etc. These Canadian technologies will lower the cost of climate compliance in India, Pakistan, China, Brazil, Indonesia, and elsewhere. Those exports are how Canada really moves the needle on global emissions. And since we're exporting all that technology, we also happen to make lots of money. What's not to like?

But that do-nothing-in-Canada argument doesn't negate the need to deploy those solutions here at home. Leaving aside for now the moral vacuity of that position, without a domestic market, we won't capture the much larger global prize. The first thing a Chilean (or Pakistani, or Brazilian, or . . .) partner will demand is, "Show me that it's working at home." If you can't do that, there's no deal.

For those on the right who deplore the idea of backing cleantech winners to gain global market share, who see this kind of public interference as a kind of far-left communism in disguise, I'd make two critical points. First, no industry of any scale, from automotive to oil and gas to aerospace, exists today without having had the benefit of strong and strategic public sector support at critical moments in its evolution. Second, there are already structural supports for existing industries. Look no further than the vast majority of EDC's and BDC's business; EDC provides nearly $10 billion in capital support to the oil and gas sector. Cleantech is like any other part of the industrialized economy with two exceptions: it's a relative newcomer, and it's working to solve existential risk.

Launching cleantech winners onto the global superhighway is highly effective and efficient. Public support within one country is levered by global adoption of more cost-efficient and high-performing climate solutions. Since it's predicated mainly on arguments related to economic gain, rather than the good of the environment, it would have minimal political cost. It may not be politically robust, however, as new governments often love to axe favorite programs of the old one just to make a point.

ARCTERN'S VIEW: SYSTEMIC INNOVATION

My own cleantech venture fund faces all of the risks and market gaps outlined above. We have intentionally built our fund in the absence of these initiatives outlined in this book on the assumption they'll not arrive in time to be of any help. As a result, we developed an investment thesis as independent as possible from the current major barriers to the cleantech market. That's possible only because we have the luxury of being a small fund (and hence can be very selective about the risks we take on) and have the benefit of hindsight.

Back in the early 2000s, there were hundreds of venture firms that listed cleantech as a priority. The big, famous Silicon Valley funds like Khosla Ventures, Sequoia Capital, and VantagePoint all put money into play. At the time, I was dazzled by their stated desire to disrupt energy systems with innovation. Here were the firms who tackled the giants of telecommunications, ready to focus on that most wicked of problems. I was particularly enamored of Vinod Khosla, one of the earliest high-powered businesspeople to speak publicly, bluntly, and openly about climate risk.

But by the time ArcTern opened its doors in 2012, Silicon Valley had their collective head handed to them on a platter. Khosla had some spectacular blowups. As a result, venture capital fled cleantech. There are maybe a half-dozen venture funds left actively investing in the sector. As I've tried to make clear, cleantech is not like IT. And you can't duplicate an investment approach that worked in IT no matter how much success, fame, and fortune that approach delivered. The irony that Khosla, a leading climate champion, inadvertently did more to destroy the financial market's goodwill for cleantech than almost anyone is especially bitter. It's essential to pick out, in retrospect, key lessons from the early Silicon Valley cleantech debacle.

A Valley VC will tell you they invest in the "team first, technology second." Which makes sense when it comes to a highly flexible, fast-moving product cycle as there is in IT. Build a minimum viable product (MVP), throw it into the market, see how people react, rebuild it in response, and the cycle continues. In IT, you pivot, pivot, pivot to ensure a product/market fit. It takes a great team to do that. But in cleantech, you're committed to a particular thermochemical pathway or power electronics system architecture — not so easy to redesign and pivot. It's more about hard science than coding, big wires not little ones, large thermal plants not data centers. If you get the tech wrong, you're toast, no matter how bright the team. Hence, at ArcTern, we look to technology first and team second. Of course, you need world-class management, but it's easier to build that team around the right technology than the opposite.

Next, we respect the profound conservatism of the utility sector. It's a lot harder to feed innovation into energy systems than communication networks. Energy infrastructure and tech risk are like oil and water. Utilities are allergic to tech risk. A utility executive's job is to keep the lights on, not play with new technology. They need equipment that operates flawlessly for decades. Same with bankers — to underwrite debt for an energy project, they need to know the technology will work for the term of the debt. A typical energy asset operates under a multi-decade power purchase agreement (PPA). The Valley tried to sell shiny new machines into an industry closed to innovation.

As a result, ArcTern focuses on what we call systemic innovation. That means investing in technologies that rely on known materials, well-understood engineering techniques, off-the-shelf equipment, and standard manufacturing techniques. We look for companies that put known things together in novel ways for a new and better result. The innovation is at the systems level, not the component level. That lowers the tech risk those bankers and utilities perceive. If each component of a system is familiar, with well-understood lifetime performance characteristics, then presumably the risk of the system as a whole is minimal. An example will bring this to life.

I flew to San Francisco a few years back to have what I thought would be a collegial discussion with one of the more famous Valley firms. I was there to compare notes on investments we'd each made in the same subsector of energy storage. Both Hydrostor and a competitor were startups building energy systems that stored large amounts of energy in the form of compressed air.[xxxvii] Hydrostor's system put huge volumes

---

xxxvii Compressed air holds potential energy, like in a basketball, car tire, or balloon. The violence of the pop, if pierced, is direct evidence of that energy.

of moderately pressurized air under water. Their competitor put small amounts of very high-pressure air in the equivalent of scuba tanks.

I began the meeting by describing how Hydrostor leveraged pre-existing engineering practices and equipment. We'd start by mining a four-hundred-meter shaft, at the bottom of which we'd dig out a large cavern the size of a football field. All this using standard mining techniques. We then fill the cavern with water from a nearby source. Two standpipes connect the cavern to the surface, one for air and one for water. To store energy, industry-standard equipment on the surface — like General Electric compressors — fills the cavern with air, and the column of water keeps that air under constant pressure. When you want energy back, the system runs in reverse: water flows down one standpipe to push the air up the other; the rushing air spins an industry-standard turbine to make electricity.[xxxviii] Bottom line: while it may sound a bit crazy, it's all standard engineering and well-understood equipment put together in a new way. We still have intellectual property, but it's not at the component level.

My colleague literally rolled his eyes: "We're trying to solve difficult problems here, Tom!" Our humble approach was summarily dismissed. He spoke at length about how they'd reinvented the art of the compressor. Every component had been reengineered at a cost of more than $100 million. The improvements they sought in thermodynamic efficiencies were elusive, he said, but possible with a team full of PhDs given the right tools and enough capital. That complexity was compounded by the massive air pressures they had to accommodate, since their business model was predicated on stuffing huge amounts of energy in confined urban environments. They had to rework the material science of the tanks for safety, since the metal fatigued with daily cycling. I agreed the problem they were trying to solve was a hard one. I wondered quietly, though, why the breakthrough energy the markets needed would come

---

xxxviii There are details, as you might imagine. For example, Hydrostor invented some subtle methods of grabbing and storing heat to increase efficiency. Compressing air generates a lot of it. Heat exchangers capture that heat, and later release it into the expanding air.

from tinkering with a technology General Electric had been building for decades!

Notice the reason our solution was dismissed: it wasn't complex enough. We weren't inventing something super-hard-to-figure-out. No question there were (and are) very bright people at that fund and the competing energy storage company. But sometimes more smarts isn't the answer — not if you've asked the wrong question. In this case, the right question is: What do my customers like? New varieties of old-school stuff will beat the newest-of-the-new for utilities every time. The role of technology in energy infrastructure is different than in industries like IT or communications.

Morgan Solar is another example of systemic innovation. Morgan is an optics company whose founder invented new ways to move, bend, and deliver light using widely available bog-standard materials. At first, they wanted to use those optics to reinvent the solar panel. That meant raising enough capital to build a factory, convince bankers the millions of panels produced are worthy of long-term debt, etc. With that approach, they faced all the barriers to the solar highway we saw earlier. Instead, they found a way to dance with existing solar manufacturers. Rather than compete with Chinese solar giants, they would license to them a Saran Wrap–thin optical sheet that reduces by half the amount of silicon required in a standard panel. Even better, that sheet is easily inserted into standard production lines. Known materials, known manufacturing techniques — not new materials and new production lines.

One of the critical questions any cleantech VC has to ask is: How can our relatively small amount of capital — tens of millions at most — have an effect on infrastructure that costs at least an order of magnitude more? Systemic innovation doesn't just address a conservative customer's phobia of deep technology risk, but also comes with its own mini on-ramp. By using materials and equipment already part of existing industries, we don't need to build a billion-dollar project to show a technology works. We build something smaller made of components with larger cousins. A small GE compressor in a demonstration project does the trick; if you want something bigger, it's in the GE catalogue! Make the new solar panel on a single solar manufacturing line; if you want more, retrofit more lines.

ArcTern tries to avoid the pitfalls and market gaps identified in this chapter. We can't do it entirely. And we're a tiny fund, occupying a small corner of the energy landscape. There remains a need to invest in precisely the technologies we avoid — new materials for solar, new battery chemistries, disruptive new machinery, large-scale biochemical plants, etc. Nothing about ArcTern's thesis negates the need for the kinds of supports outlined in this chapter. Indeed, the strict limits under which we operate provide two practical lessons, both of which corroborate the supportive policy outlined in this chapter. First, public investment in R&D is necessary, but insufficient: necessary, because venture funds don't invent anything (we commercialize what's already invented); insufficient, because more, much more, is required to support deep innovations getting to market at a scale relevant to an effective Climate Capitalism, which can't happen with private capital acting alone. We will need that deep tech.

But most importantly, nothing I see on the cleantech horizon[xxxix] — no matter how low its perceived technology risk, promising its economics, or clever its management team — is scaling anywhere near fast enough to slow climate disruption. As I tell my partners: I may be optimistic on cleantech as an investor, but I'm deeply pessimistic on climate as a human. Public policies to accelerate our transition to a low-carbon economy are badly needed no matter how much small, nimble cleantech companies find ways to grow in their absence.

THE BEST OF PUBLIC AND PRIVATE: BANKS AND BONDS

That's an awful lot of market gaps to fill! Change the attitude of CFOs to see energy costs as being of strategic interest; fund large demonstrations of EGS and NGN and other unique large-scale technologies; generate scale of production for emerging technologies and debt for their customers; accelerate cleantech champions into global markets — and that's just the start. The filler for these gaps is always predicated on public support of some kind. The leap from innovation to market acceptance cannot

---

xxxix We've looked at more than five hundred cleantech companies across the globe. Take my word for it: there are no magic bullets out there!

be achieved without the catalyst of public funding or public policy, or both. This is not unique to cleantech, although the problem cleantech is solving is uniquely urgent! Automotive took off with interstate highway systems, nuclear needed the military, as did aerospace and even the internet. What is to one person a subsidy (typically the perception on the right), may be to another (on the left) a good long-term investment. Let's be clear: it's both. While the gap to market exists and requires public intervention, it is indeed a subsidy. While the planet continues to heat, it's clearly a critical investment.

Calling the challenge to cleantech a market gap — rather than a failure to commercialize — indicates it's a temporary aberration of the natural expression of self-interest that drives markets. This isn't to say there aren't market failures. There are. A price on carbon fixes a big one. And some initiatives like fuel efficiency standards or building codes are best left to a regulatory framework. We're after something different here. Like priming a pump that operates on its own once it's up and running. Or accelerating what would happen on a longer time frame anyway. Across most of the economy. A green bank to Climate Capitalism is what a good coach is to a hockey team: the players may be fine on their own, but if winning is important, you need that coach. And winning here means moving fast enough to beat down that climate hear.

Clearly a conservative's nightmare! Massive government bureaucracies, churning out hard-earned taxpayer dollars. No market rigor to guide decisions. Surely such a smorgasbord of subsidies will end up with massive waste and even fraud. Right? Perhaps. Ontario's feed-in tariff comes to mind, a policy in which solar subsidies were insulated from the precipitous drop in the cost of solar panels, resulting in huge overpayments to developers. Equally, there have been successes — the U.S. Department of Energy's loan guarantee program, for example. Or Sustainable Development Technology Canada's (SDTC) renowned technology scale-up funding.

But conservative concerns have merit, particularly as we seek to scale up the pace and scope of public support. It's critical to bring market rigor, transparency, incentive structures, and accountability to public subsidies. While that sounds like a contradiction in terms, it's not. Most of the gaps identified in this chapter are temporary blocks to flows of private capital. Over the long-term, each will slowly disappear under a clear and long-term carbon pricing signal. But we're so late to the climate game, in the interim, we need to short circuit the market into early action. And rather than hand all this activity to a government ministry — which on the best of days is only marginally sensitive to market dynamics — we might design a hybrid of public and private.

Hybrid Governance: Green Banks

An independent green bank is a publicly mandated institution designed to accelerate financial flows to low-carbon infrastructure yet is responsive to market conditions. Think of it as a vehicle designed to unblock capital flows, whose interventions in the market are by design temporary. There are precedents. New York State and the U.K. (among others) established green banks to accelerate their own low-carbon transitions. The U.K.'s green bank supercharged their world-leading offshore wind industry, which is expected to provide nearly 30 percent of their electricity by 2030. In New York, it's a broad mix including lots of solar and energy efficiency. Canada and our provinces would do well to follow their lead.

What kinds of activity might a green bank undertake? Certainly, to close all of the gaps we've identified thus far. The range of activities is open-ended, as are the financial tools it can use. Both vary with market conditions. And that's the point: no government ministry or fund can react to market dynamics the way an arm's length institution can. Think of a green bank as a kind of sensor embedded deep inside local markets, able to detect and respond to their signals in a way no government ministry ever could. It couples information from the market with a public mandate to accelerate clean energy infrastructure and reduce emissions. Entrepreneurs bring technical innovation. A green bank brings financial innovation.

It limits its role to market formation, providing capital in ways that catalyze private sector actors. Once they're present, it moves on. It cuts through the grooves, habits, and barriers developed over many decades of our economy's evolution. It's willing to take risks the private sector won't due to human and institutional bias. It does so to fulfill its public mandate: accelerate climate solutions into the market. It works to recoup capital it deploys, which is recycled back into the market. Its money isn't free, but it is provided below-market rates. Some examples will bring this idea to life.

Pension funds are the single biggest source of long-term capital. They'll trade risk for long-term stable income all day long. Perfect for energy retrofits like Planet Traveler across the entire economy. But pension funds won't invest without massive scale (the number has to begin with a "b"!). Projects like Planet Traveler are tiny in comparison, and so face high capital and transaction costs. A green bank can aggregate, or warehouse, those assets. That provides scale and eliminates liquidity risk (the risk that an investor can't resell the asset). The aggregated assets are then sold to the private sector, like pension funds, in suitably sized chunks. Once the relevant scale of capital starts flowing, the green bank steps back to let private actors take over. It might need to do something similar for each subsector, acting as an industry-by-industry financial warehouse for energy retrofits deploying best-in-class technology relevant to that industry.

A green bank might provide early risk capital in the form of first-loss provisions or loan guarantees for those two underfunded technologies of the big three — NGN and EGS — or any other large CapEx first-of-its-kind cleantech projects. How do we use market rigor to manage that risk? By matching long-term reward with the form of risk it takes. For example, tie support for an EGS consortium to long-term license fees on future deployment, paid back to the green bank as reward for taking on that early risk. Or tie early first-loss capital loaned to local, first-of-its-kind next-generation biofuels facilities with upside on future global deployments — this could be an option to invest equity at a predetermined value or the right to finance future projects at a predetermined (above-market) interest rate. In both of these examples, the rights the green bank obtains can be sold to the private sector once the market is established, returning capital to its coffers.

There are lots of other ways for a green bank to play. It can build the on-ramps identified in "Freeways and On-ramps" with low-cost capital for manufacturing capacity to supply early global contracts. Coupled with first-loss subordinated debt for those customers, it unlocks commercial bank activity for other projects coming from the same factory as the technology gets to scale. It might fulfill Own the Podium by matching private funds on hard-to-find project equity for large-scale buildouts of newer technologies, triggering that move when the company gets a large contract with a tier one customer. By doing the technical diligence a traditional bank won't, it can sell well-priced performance insurance. That translates to security for private lenders who worry the project might not spin cash for the long-term. Or, it can extend the maturity of available debt to match a project's lifetime. Or, deal in retail financial products that get small businesses and homeowners involved. And so on.

There's lots of precedent. Green banks are, in fact, a new twist on an old theme — arm's-length institutions that apply the rigor and creativity of private markets to public mandates. SDTC is a world-class early-stage clean technology funder and a big part of Canada's early lead in cleantech. EDC has always been a critical partner to Canadian exporters. And the BDC has long provided liquidity to small- and medium-sized businesses, often in the absence of the big five commercial banks.

Critical to success of any green bank is how the mandate is defined. Incentive structures for investment managers need to be linked to the long-term outcomes most valued (Key Performance Indicators, KPI, in corporate-speak). The top KPI is obviously total emissions reduction, with a close second being the pace at which supported sectors convert from public to private support. KPIs should not be linked to short-term return on capital (the Achilles heel of the private markets) but instead to the way risk is linked to environmental performance: risk is increased commensurate with emission reduction potential. A few "swing for the fences" investments are balanced by warehousing functions — low risk, but high transaction costs.

Private markets are not perfect, and sometimes they need a push, particularly when it comes to huge investments in deeply innovative new technology. Capital often overstates the risk of the unfamiliar while understating the risk of embedded behavior. The deep conservatism of capital markets means security and a steady hand, desirable in many ways, but not the kind of thinking that drives the fundamental and urgent economic restructuring that climate risk calls for. The pump for our fire hose needs priming.

Hybrid Financing: Green Bonds

Nothing in life is free, and even if the green bank succeeds spectacularly in igniting private sector activity, the seed money has to come from somewhere. The obvious choice is government. Governments raise money in two ways, taxes and debt. Carbon taxes are an obvious choice, but it seems politically astute at this early stage to render them revenue neutral. Jatin Nathwani of the University of Waterloo recently suggested a 2-percent increase in sales tax (GST in Canada) to fund low-carbon infrastructure. That would generate about $20 billion a year, perhaps ten times that in the United States. But resistance to general taxation is even higher than to carbon pricing. Government debt is straightforward, but it needs to be paid back with future tax revenue. It just kicks the can down the road.

There is a more intriguing possibility. Well, actually, there are two. It's worth pointing out here that governments printed money to save the financial sector starting in 2008; Massive quantities of it through their central banks. Quantitative easing was the polite term. But it really was just printing money. If we can do it to save banks, surely we can do it to save the planet. Banks are important, but not yet on the same tier as a stable world. But that suggestion,[50] while pragmatic, will likely meet a maelstrom of criticism from the right. We avoid that maelstrom with another suggestion.

During World War I and World War II, many countries issued bonds (known as war bonds or Victory Bonds, at least to the victors!) to support their respective war efforts. These were a variant on standard-issue government-backed bonds, but themed in a way to excite and motivate a patriotic general public. A variant on the war bond might be a climate bond, but with a twist. Back in 2009, I and a group of Action Canada colleagues proposed a policy initiative called the Canada Green Bond. On the surface, it looks like a Canada Savings Bond: funded by individual citizens and issued by the federal government. But for the Canada

Green Bond, the capital raised is managed by the private sector, which is motivated to recycle that capital — again and again — as low-cost debt across the low-carbon sector. The big difference is how the money sits on the government's books: it's not debt, but a contingent liability.[xl]

To be clear: this is not free money. It is a subsidy in the form of a lower cost of debt servicing and larger quantities of capital than would normally be provided by the private sector. Any money not paid back is lost, as is overhead. All of which has to be covered by the taxpayer, the ultimate guarantor of the bond. Hence, the contingent liability. But our calculations showed it is — *by far* — the lowest-cost (to government) way to fund an acceleration to a low-carbon economy. Why? The unique pairing of public backing and private management combines low-cost debt with an aggressive and market-oriented mandate to generate a return. It's not free — no subsidy is — but the cost to government is a tiny fraction of the capital deployed.

Don't confuse the Canada Green Bond with the green bonds that are now standard-issue corporate debt. While the term now refers to almost any corporate or public bond issued with the intent to underwrite a project with positive environmental outcomes — from Ontario's subways to a corporate wind farm — the original idea was a public/private partnership in Europe. The Europeans issued a Climate Awareness Bond in 2007 to support their sustainable energy sector, and the bond sold out in three months. The World Bank followed suit in 2008.

Both of those initiatives were policy tools that brought together two ideas: the public's desire to invest in clean energy projects coupled with a need to bring down the cost of capital for those projects. At that time, the perceived technology risk associated with wind and solar farms was too high to enable low-cost private sector debt, which kept the cost of clean power artificially high. And while the public was willing to support these projects, there were few ways for retail investors to play. So bonds

---

xl   Only some portion of the total amount will be required to be repaid by government, since the money is *invested* rather than *spent*. As an investment, it generates an asset that has value. In this case, the value is less than the full amount, but it's clearly not zero. Our calculations indicated a contingency factor of 75 to 90 percent.

were issued to the public with the principal backed by the balance sheets of national governments. The result was a flow of low-cost debt from an eager (but risk-averse) public to renewable energy developers.

The Canada Green Bond is a variant on these ideas. Canadians buy a government-backed bond (just like a Canada Savings Bond), and the funds raised are used to accelerate renewable energy production by providing low-cost debt to energy producers who choose renewable production methods. At its core, it's that simple. Since we first proposed the idea in 2009, the perceived risk of wind and solar have dropped considerably (validating that it was overstated at the time), and so the target projects of the Canada Green Bond would shift to support the rollout of next-generation technologies. In precisely those ways associated with a green bank.

There are two sides to a bond issue: selling the bond and deploying the funds. Public engagement to raise the money is the most visible portion. A Canada Green Bond sold to retail investors has the benefit of informing and exciting the citizenry. It provides an answer to those people asking, "What can I do?" with a better answer than "Change your lightbulbs to LEDs!" Green bonds are a popular idea with the public: in polls we conducted, more than 80 percent of Canadians supported it. They should be popular with politicians, too, giving them a tool to engage the public in a positive way on an issue normally associated with tax increases, policy imposition, and doom-and-gloom environmental outlooks. Indeed, I think one of the most valuable outcomes of a national green bond is the chance to flip the public conversation to one of engagement, education, and participation.

Raising money from the public is visible, but not all that efficient. The real money would come from institutional investors who always place some portion of their funds in the safety of government bonds. To dial up the amount raised, the feds just need to add a tiny premium above the national borrowing rate. That could be a simple uptick in interest, or it could be a more creative premium linked to outcomes like absolute carbon reductions. Today's capital markets are starved of decent returns that are safe. Any national-backed bond with a small premium will stick out like a flashing light and attract significant capital flows.

The financial details are the other side of any bond issue. The unique part of our proposal was a new form of public-private partnership that maximizes the policy's effectiveness and minimizes its cost. The nuts and bolts? The government backs the bond and provides a mandate to the private sector defining desired outcomes. This can be done under the purview and oversight of a green bank. Fund managers bid on the right to manage some portion of the funds. The selected fund managers then use all their creative talents to deploy the capital to meet that portion of the green bank mandate while minimizing cost to government. One group handles energy efficiency warehousing. Another, scales to manufacturing. Yet another handles first-loss on large CapEx projects.

Why private sector management? The government shoulders the risk, so why separate risk from management? First off, no one likes the idea of the government picking winners. More importantly, we want to leverage the creativity and efficiency of the private sector by offering the right financial incentives to deliver real financial discipline. Link the money managers' upside and bonuses to precisely those outcomes we want, like the total cost to government (which backed the bond) per ton of emissions reduction. In other words, use the potentially ruthless attitude so easily adopted by private money managers to reduce risk. What's the biggest risk? Deploying money that isn't repaid. That's the cost to taxpayers — stranded loans.

This strategy is certainly interventionist — driven by public policy and backed by government. And, it does have a cost to the taxpayer. But because it leverages the behavior, insight, and efficiency of markets to accelerate a public good — our transition to a low-carbon economy — it's less a leftist government scheme and rather a truly centrist idea that opens up a better way forward than anything the right or left could or would advocate on their own.

Imagine the money is provided as debt to build a high-voltage direct current (HVDC)[xli] line to Hudson's Bay. That opens up those huge wind resources to developers, who can then pump massive amounts of clean

---

xli   A high voltage direct current transmission line can move enormous amounts of power very efficiently, over very long distances.

power through the North American grid. Wind developers build offshore wind projects, rent access on the HVDC line to get the power to market, selling it to our American cousins south of the border. Now imagine that the managers of that publicly backed HVDC line do a lousy job maximizing revenue or get snookered by the private wind developers in complex offtake contracts. Or there is a temporary glut of power, or a deep American recession, that forces the wind developers into bankruptcy.

In any of these cases, the hard assets (the HVDC line itself) have real value. As do the offshore wind turbines. So, too, do the power purchase agreement contracts themselves, predicated on long-term relationships with utilities south of the border. Now imagine the government trying to repossess these assets, and reselling, redeveloping, or renegotiating their contractual provisions. Or, for that matter, doing the right kind of due diligence in the first place on the HVDC management team. None of these activities sits well in government ministries, but these are things at which a motivated private sector excels. The left excoriates the ruthlessness of markets, but it's precisely that discipline that avoids white elephants.

Time and again, the public sector has demonstrated substantive inability to engage in precisely the kind of aggressive, self-interested (and enriching) behavior that drives market discipline. The Trans Mountain Pipeline debacle comes to mind.

Another example among many: when the Canadian government bailed out Bombardier (for the umpteenth time) they showed little ability to get a good deal for the taxpayer. They seemed more motivated by the short-term benefit of announcing saved jobs, and very likely succumbed to behind-the-scenes political arm-twisting. Bombardier got a great deal, the taxpayers a lousy one. Any private sector actor in that situation would have been more than willing and able to push back and, at the very least, would have both taken control of Bombardier and generated better financial upside by getting preferred shares (for example). The point here is a general one: only the private sector, unconcerned with optics and motivated by self-interest, can negotiate a deal like that. But without policy to create low-carbon guide rails, all that expertise is lost to Climate Capitalism.

What's the cost of green bonds to government? The main cost is loan defaults, which the fund manager is motivated and able to minimize. Other factors are asset recovery rates, management fees, whether the borrower has to put up matching funds, etc. We did the math. The worst case for loan defaults is measured as the difference in lending rates between what the market is willing to provide and the subsidized rate the green bonds program provides. That's because the market rate *includes the risk of loan default.* The better cases assume the market overstates that risk. Today, that spread, and the estimated cost to government, is something like 10 to 15 percent.

In the three scenarios we ran, costs ranged from $1 to $13 per ton of avoided emissions. The lower end of that range corresponds to a realistic scenario. The upper end to the sort of worst-case-the-sky-is-falling analysis required by the folks at the Ministry of Finance. This looks a lot like the U.S. DoE program. And it is. But the money comes from investors, not taxpayers. The taxpayer contributes only the final cost in the form of losses. In the U.S. DoE case, recall that was only 3 percent. This points to a rosier scenario of much less than $10 per ton.

That's cheap! The federal government has proposed carbon pricing starting at $20 per ton starting in 2019, increasing to $50 by 2022. The social cost, what unchecked carbon emissions will actually cost us all, is much higher than that. Why is our proposal so cheap? It effectively limits the government's exposure to paying only for those loans that are defaulted, and the loan agreements and risk-mitigation efforts sit squarely in the hands of the private sector. And each dollar of cost to the government is multiplied many times into actual capital deployed to produce renewable energy.

The green bond idea is flexible. Technology choices, lending rates — all the details of implementation — can change according to market conditions. That's the role of the green bank: to be creative in how that money is best used. It targets a much broader range of companies than either tax credits (to benefit from these you need to be profitable) or fixed subsidies. And it's temporary; this subsidy will quite naturally drop away as costs of carbon emission compliance increase (slowly, over time) or as commercial banks indicate their willingness to lend at similar rates.

In Canada, the railroad was built on innovative financing mechanisms. More than a transport corridor, it was a nation-building project that sparked the imagination of the country while simultaneously providing an economic backbone to last a century or more. Imagine the sort of nation-building project green bonds could support: high-efficiency DC lines up to Hudson's Bay, opening the north up to wind developers. A nationwide building retrofit program that employs tens of thousands of well-paid tradespersons. All of the initiatives that accelerate Climate Capitalism can be funded this way. Build the bond, and the buyers will come. Build the bond, and help build a new, greener Canada — a renewable energy superpower.

Climate Capitalism is about much more than nation-building, however, even if most of the action happens at the national level. It's really about coordinating an accelerated response across the globe. How we might move from the level of independent (often bickering) sovereign political entities to something larger is the subject to which we now turn.

CHAPTER SIX

FROM PEOPLE TO THE PLANET – HOW SOVEREIGN IS SOVEREIGN?

As we move into the twenty-first century, one thing is increasingly clear: our world is a shared one. From global trade to flows of refugees, we are connected through finance, empathy, politics, and more. And, of course, our ever-more-angry atmosphere is common to us all. The days of isolation from one another are well and truly over. To some, this brings great anxiety, both cultural and economic. Trump's win was predicated largely[i] on tapping the fears of voters who were concerned for both their economic well-being in a world that was becoming highly competitive, as well as their identity in a world that was beginning to look a lot different. To others, this global interconnectedness brings opportunity and social rejuvenation. Regardless of one's views of the evolving human cultural condition, boundaries are not as clear as they once were. That is the case for climate risk as never before. Climate disruption knows no borders and respects no walls.

To have sovereignty is to be independent — to make one's own rules in one's own interest. As climate disruption begins to bite, the

---

i   And with more than a little help from Putin, of course.

very notion of sovereignty will shift. Canada's intranational squabbles on climate are over relative shares of the economic burden of emissions reduction. We made promises as a nation, but the work is largely borne province-by-province as economic conditions indicate. We can go ever-lower in a chain of sovereignty to the atomic entity that is the (somewhat) sovereign person and ask, "What is my climate obligation compared to my neighbor's? To others halfway around the world?" Or we can go ever higher to the ultimate sovereign: an Earth that goes about its business on a time frame nearly inconceivably long and cares not a whit for human affairs. In this chapter, we move through links of sovereignty: national promises shared by subnational economies; international agreements to link national promises to global targets; and carbon pricing to link our economies to values inherent in the earth itself.

COUNTIES AND COUNTRIES, STATES, AND PROVINCES

We've seen interminable arguments over how much each country should contribute in the global fight to reduce emissions. That's hardly surprising. It's at the nation-state level that Paris targets are set, which corresponds to where laws might be enacted to meet those targets. Similar arguments erupt within nations. In the U.S., coal states fight for their right to keep digging up the black stuff. In Canada, the showdown is caricatured as a fight between Alberta and the rest of Canada. While they have less merit, these subnational arguments bring into sharp focus the pragmatic and moral stakes at play. Let's take the Canadian example, which often comes down to fights over pipelines to deliver Albertan heavy oil to market.

There is no more controversial energy project in Canada than the Trans Mountain Pipeline. But absent the Paris Agreement, the furor over the Trans Mountain was overblown. Despite escalating rhetoric of premiers Horgan in BC and Notley in Alberta, jurisdiction falls squarely on the federal government. By following an existing route to a bus-tling port, it's one of the least intrusive pipelines on the books. Worries about spills, en route and in ocean waters, are real but manageable. And if investors want to bet on long-term, high-carbon infrastructure in a carbon-constrained world, they should be free to do so — if they also agree to accept the risk of having those assets stranded when the economy shifts to low-carbon.

But now things are so crazy, the Canadian government bought the project and committed to doubling its capacity. In doing so, they relieved American investors of a crappy deal and took on both financial *and* legislative responsibility for it. The political calculus is obvious (save Alberta votes), if cynical. The feds are now pushing and pulling on emissions at the same time: all their cleantech funding efforts were swamped by that *one* risky, multi-billion-dollar, high-carbon project. The Canadian government now plays all sides in the carbon game. Oh, that pesky Paris accord.

Canada faces a shrinking emissions pie. Our Paris pledge — to reduce emissions 30 percent by 2030 and 80 percent by 2050 — is a moral commitment shared by us all. When a single province, industry, or company increases emissions, a heavier burden falls on all the others. Behind the Trans Mountain debacle — squabbling over access to tidewater, rights of way, and safety — lies the real fight: struggling to achieve our lower emissions targets as a nation. It's a long-overdue conversation about shared climate burden.

Trans Mountain is a tangible expression of fossil fuels' continued dominance. For many protestors, it signals expansion of Alberta's heavy-oil production, already the largest industrial source of emissions in the country. Production growth depends mainly on oil prices, of course. Below $60 per barrel, there will be little increase in output. But the federal government's recent financial backing for Trans Mountain, in addition to increased regulatory certainty, supports the idea that production expansion in the oil patch and shared climate commitments go hand-in-hand. They cannot.

Imagine a boat crossing a shallow river, rife with rocks. To be safe, the boat's draft must be reduced so it sits higher in the water. Each of ten occupants is expected to play their part and ditch luggage. One occupant insists they uniquely have the right to take on more weight, not less. The captain even buys them a bigger backpack. Every extra kilo increases the sacrifice of others and eliminates all sense of fair play.

The math is cruel. Alberta's emissions from the heavy-oil industry are seventy megatons[ii] today. Alberta's current cap allows an increase by almost half, to 100 megatons. Were Canada to meet our 2050 commitment, by then Alberta's emissions would represent more than two-thirds of the entire country's emissions. For less than one-twentieth of the economy. The largest beneficiary of expanded production? Notorious climate skeptics — the Koch brothers are the single-largest bitumen lease holder in Alberta. A strange irony given it's become standard fare in Canada to accuse environmental groups of being somehow disingenuous or even traitorous because a small portion of their funding is non-Canadian (as if foreign funding is somehow a bad thing).

The rest of Canada cannot make up the difference. Every other industry and province needs near-zero emissions to make room for increased bitumen production in Alberta. Why should Ontario's manufacturing, New Brunswick's farming, or Quebec's pulp-and-paper industries work so hard to have it all undone by a few new in situ bitumen mines? It's reasonable to ask of our shrinking emissions pie: What size emissions are allowable to any company, industry, or province, and for whose benefit? The same can be asked of any emissions-intensive industry, such as BC's proposed liquefied natural gas exports.

Maybe technology combined with a carbon price will enable emissions-free bitumen production. Modular nuclear units powering Cenovus's in situ mining, for example, or solvent-based extraction in lieu of heat. That would certainly lessen Alberta's slice of the pie and open up more emissions space for other Canadian industry. We've seen impressive advances in innovation in heavy-oil production thus far, and the per-barrel emissions have come down — but never have they led to a decrease in absolute emissions from the industry as a whole. If industry believed technology could solve their emissions problem, there'd be no need for a cap 50 percent higher than today's emissions!

---

ii   Recent studies indicate current emissions may be vastly understated, which means we might be near or even past that level now. See John Liggio, et al., "Measured Canadian oil sands CO_2 emissions are higher than estimates made using internationally recommended methods," *Nature Communications* 10 (April 2019): 1863.

Arguments that Alberta, as a resource-based economy, has a more difficult job reducing emissions than other jurisdictions ring hollow. By definition, emission reductions are not the same everywhere; not everyone has to hit the same number. They're measured from baselines, which take into account existing vagaries of geography, reliance on resource extraction, even how many citizens choose to drive F-150s or Ford Broncos to hockey practice.

That's why subnational arguments of this type have less merit than those between nation states. Provinces certainly have different economies — that's reflected in their respective baselines — but they have much more aligned economic outlooks and histories than different countries. Alberta compared to Ontario is very different from India compared to Canada. Developing countries differ from developed countries in ways that are not reflected in the baselines against which reductions are measured. Developing nations do not have the same historical emissions; they have much less public wealth to deploy in the climate fight; the majority of their citizens have yet to enjoy the economic growth that led to today's middle class; and so on.

Intranational squabbling in developed nations can sometimes get embarrassing. Some radical Albertans have even argued they'd be better off outside the Canadian family, free of the constraints of having to meet emissions targets agreed to by a federal government. Most obviously, that wouldn't get a pipeline built. Worse, as a country, Alberta's per capita emissions would be the highest in the world — four times those of Saudi Arabia. I doubt young, savvy Albertans are willing to wear the moral shame of withholding climate commitments entirely. Without the rest of Canada as a buffer, Alberta's "national" reductions would by necessity be much sharper.

We've built pipelines before, and building infrastructure always makes someone angry. All energy projects bring risk, but that risk is manageable. These issues are not new. They're proxies for the real heat behind projects like Trans Mountain. We're finally beginning a long, difficult national discussion about who gets to emit, how much, to whose benefit, and at what cost. A uniquely Canadian family squabble. And one we need to show the world we can resolve. Each of our provincial starting points, whether you're from Newfoundland, Quebec, or Alberta, is one of relative ease compared to our friends in India, Pakistan, or Venezuela.

SOVEREIGNTY AND TRADE AGREEMENTS: A NEW WTO CLIMATE CONSENSUS?

For decades, political leaders of all stripes endorsed increased global trade as a means to increase wealth, raise living standards, and even reduce the possibility of conflict. Globalization was political orthodoxy across the world, but those cozy assumptions got upended as populist voices got louder. First, the U.K. rejected the European Union — a political extension of an economic union — in the Brexit referendum. Then, the standard-bearer of free trade fell with a thump in the 2016 U.S. general election. Trump played on the fears of a citizenry who felt left behind by policies seen to benefit the elite. The far left has long fought against rules seen to prioritize the rights of capital over people. Trump's surprise win launched a new economic nationalism from the right.

But the enemy isn't globalization, or even global trade itself — both of which have been around since people got in boats. The enemy is economic and political insecurity. Only insofar as people feel disempowered or threatened do they target trade deals or want to put up walls. When Trump ripped up the Trans-Pacific Partnership (TPP) and then started a trade war with China, he signaled to his supporters that the U.S. would regain stewardship of its own economy. Yet a reduction in trade will in no way be a net benefit to the American worker or the U.S. economy as a whole. Without access to cheap goods from abroad, American living standards will fall. Absent ongoing integration into the global marketplace, the U.S. economy itself will falter.

But just because populist voices can be mistaken in rejecting the net benefits of global trade, that doesn't mean these agreements aren't subject to legitimate criticism. Trade agreements became a target because they're seen, often correctly, to skip over the rights and interests of individual citizens in favor of the investor class. Free trade is like a club. To join, you agree to obey the rules. That means giving up some degree of sovereignty to supranational institutions like the World Trade Organization, which impose rules on national governments and decide (often behind closed doors) what is and isn't allowed. The whole point of trade agreements is to put the rules out of reach of legislators so business can rely on one set of unchanging rules everywhere.

Under international trade agreements, a corporation might prevent public support for an emerging competitor or sue over "buy local" provisions when they hurt their prospects. Governments can be prevented from supporting strategically important or politically popular industries. But profits are better protected against legislative action than the environment. Because states often have to push so hard against their own citizens' legitimate concerns about the subversion of democratic to corporate rights — recall Seattle in the late 1990s — economist, philosopher, and prolific author John Gray calls these global agreements "forceful" state action.

These agreements reflect the priorities of those who negotiated them. They were created expressly to lower barriers to trade and capital. They were not created to protect the environment, empower local governments, protect workers, or enhance human rights. If present at all, these ideas were afterthoughts. Democratically elected representatives were nominally engaged in negotiations, but the terms of engagement reflected the ascendant neoconservative economic orthodoxy of the time (the Washington Consensus). Citizen groups didn't have a real seat at the table. At best, they were observers. Nothing about capitalism itself precludes a different outcome, however. A robust Climate Capitalism would ensure these agreements have embedded within them an urgent response to climate disruption. With the right adjustments, these agreements can become *tools* for Climate Capitalism.

If a man like Trump can disrupt long-standing alliances like NATO, throw punishing trade jabs at Canada under the clearly disingenuous auspices of national security, and slap China with an "own-goal" set of tariffs, then he has done one good thing: to demonstrate that a motivated, strongly backed leader can upend the norms by which the global economy and security systems have operated for years. I'm no fan of upsetting current global security agreements, nor backing away from global trade. Tearing something down is no guarantee something better gets built. But in the face of a climate emergency, it's reassuring to know the deepest, most treasured institutions can be renewed with sufficient effort.

Critics of the WTO and various trade agreements like NAFTA correctly point out that they prioritize the liberalization of trade and capital flows as an engine of economic growth over everything else. This is a legitimate source of frustration to social activists the world over. As a result, some want to see the WTO abolished. Others want reform. I fall squarely into the latter camp. Co-opting the most powerful economic institution in the world could embed climate action and accountability directly into the strongest rules governing economic activity. Countries long ago gave up sovereignty to join the WTO. Let's use that lever to force climate compliance.

Since trade agreements are now under scrutiny by both the right and left, we have an opportunity to rethink their purpose. It may take Herculean effort, but trade agreements and institutions like the WTO aren't immune to reform. Certainly, maintaining global trade and facilitating the integration of economies will remain a priority. But there's no reason these same agreements can't reflect the new priorities of an aggressive Climate Capitalism. And they have teeth. We can use them to go after carbon laggards and encourage climate action.

Instead of a "Washington Consensus," one might imagine a "Climate Consensus" as the basis for inclusion in a renewed global trading club. If countries gave up sovereignty for trade, surely the same is possible for a livable climate. A climate agreement with teeth means legally binding commitments to reduce emissions and mechanisms to punish those who don't comply, as opposed to the voluntary, nonbinding nature of the Paris Agreement. Climate activists have been waiting in vain for such an agreement for more than two decades. Skeptics have long argued it's unrealistic to think countries will give up sovereignty for climate. The United States and its dogmatically recalcitrant Congress is the prime example. Voluntary targets, we're told, are the best we'll get.

Yet every country in the modern global economy gave up sovereignty in the name of trade liberalization. The WTO prevents democratically elected governments (and a few autocrats) all over the world from passing all manner of laws, regardless of electorate support. Trade agreements from NAFTA to the recently contested Trans-Pacific Partnership (TPP) do the same thing. Free trade, the bastion of the right, is predicated on lots of rules and regulations that bind government hands. Transgressions and appropriate punishment are decided by small panels of unelected officials, often behind closed doors.

That's not necessarily a bad thing. Large international corporations want to maximize the flow of goods and capital across borders, and governments see increased trade as a means to create wealth and jobs.[iii] The purpose of free trade was to eliminate the problem of having to play by different rules in every country and minimize the risk of subjecting long-term investments to the whims of unstable governments. Trade agreements have been remarkably effective at achieving these objectives. And global trade agreements brought economies of scale and reductions in cost in several industries. One reason solar and wind dropped in price by more than 80 percent over the last few years is because trade agreements facilitated mass production and linked global supply chains.

But that's not the whole story. Even while signing trade agreements, countries have always tried to support nascent domestic industries, some more aggressively than others. China's state-owned banks provided huge amounts of below-market capital to its solar manufacturers as part of their takeover of that industry. China's market just happens to be so big no one wanted to pick a fight until Trump came along (the only potential upside of his presidency that I can see). Others are not so lucky.

India's National Solar Mission is an ambitious attempt to ramp up solar production and is a centerpiece of its COP21 climate commitments. They ramped up from nothing to more than five gigawatts in just five years. The aim is 100 gigawatts. To crank deployment and build domestic capacity, India offered subsidies to kick it all off. Given their overreliance on coal, their climate commitments, and awesome solar resources thanks

---

iii   I will not address the thorny issue here of whether or not this is the best way to promote economic self-interest and support decent jobs. Critics of free trade point to the inevitable "race to the bottom" in terms of wages, labor rights, and environmental protection. Supporters argue it leads to everyone being better off in the long run. As usual, the truth lies somewhere between the two extremes.

to their geography, this strategy makes sense. India also needs jobs, so why not link public subsidies to support domestic manufacturing? In this case, the motivation is politically pragmatic; selling climate action isn't easy,[iv] especially in the developing world, but selling jobs is. Obama seemed to support the idea of solar subsidies after visiting Indian Prime Minister Narendra Modi in 2015. The two released a joint statement "emphasizing the critical importance of expanding clean energy research, development, manufacturing, and deployment, which increases energy access and reduces greenhouse gas emissions." So far, so good.

But the U.S. had *already* launched a suit[v] against India in 2014 under the WTO. The WTO ruled against India, saying it broke the rules by favoring domestic manufacturing of those solar panels. Even worse, the ruling explicitly stated that the WTO rules override other treaty commitments, including the climate agreement signed in Paris. So antiquated WTO rules trump current climate needs, handcuffing progressive national leaders. That's absurd.

This is not an isolated incident, but a clear battleground of principle across multiple trade agreements and sectors. The draft of the TPP contains the same language as the WTO on "buy local" provisions — governments cannot favor local production. As countries around the world work to build broad constituencies of support for climate action, they need all the help they can get. There's a cost to aggressive climate action, and the ability to link local economic benefit to those costs is critical to build public support. It also creates more long-term competition as countries scale up their own production to compete for global business.

Trade agreements are not bad in and of themselves, but they currently reflect a narrow view of economic activity and its relation to other

---

iv   When Ontario tried to seed domestic solar technology production in its Green Energy Act, the same thing happened: the WTO ruled against it and companies shuttered. Ontario is a developed economy and the ruling was a tiny bump in the economic road (as annoying as it was to companies and investors who'd committed to production on the government's promise).

v   The unpleasant irony of the U.S. acting the international bully is that more than half the U.S. states have "buy local" provisions.

urgent international commitments. This is not surprising, given that was their intended purpose when they were formed decades ago — to elevate the flow of goods and capital above other considerations and provide corporations with a stable investment climate. And governments voluntarily gave up sovereignty to allow it. That creates an opportunity to reform those institutions to accommodate climate as an alternative to building new ones from scratch.

There are two ways the WTO should be reformed. At the very least, it shouldn't get in the way of countries that need to protect or nurture low-carbon industries as part of a comprehensive climate response. The WTO must make exceptions to allow for countries to protect local cleantech industries, especially when those countries are aggressively providing economic incentives to accelerate the deployment of low-carbon energy.

A more ambitious reform is to empower the WTO to enforce international compliance on emission reductions. Countries that refuse to step up on climate — who do not price carbon when their trading partners do so — would face penalties on their exports. Creating a level playing field is in the spirit of the original mandate of the WTO, and carbon tariffs on goods from non-compliant countries would prevent "carbon leakage," the inevitable race to the bottom as carbon-intensive industries try to move from countries with aggressive climate laws to those without. That way countries can pass aggressive climate legislation without having to tie themselves in knots to protect local, emission-heavy industries. That's exactly how Ontario designed its now-ditched cap-and-trade system: exemptions were doled out industry by industry to keep them competitive against global peers who face no comparable emissions constraints.

All this talk of reform may sound overly ambitious. After all, the WTO doesn't seem remotely democratic. Its decisions are often opaque and run counter to the expressed will of democratically elected governments. But that's just the nature of international agreements; giving up sovereignty is part of the game. The WTO is just a club run by its members. And its members are representatives of democratically elected governments. It's a tall order, but citizens of each member country can demand their government work to reform the WTO to turn it into a powerful tool of climate compliance.

While a reformed WTO that prioritizes climate compliance sounds like a long shot, it's a better bet than starting a new international agreement from scratch. And given the WTO's ultimate objective is to maximize economic growth — after all, that's the justification for trade liberalization in the first place — it's not a stretch to argue climate compliance is already implicitly a part of their core mandate. There is no long-term economic growth without urgent and unprecedented climate action.

A SOVEREIGN EARTH: PRIVATIZING NATURE OR VALUING NATURAL CAPITAL?

We are all answerable to a functioning ecosystem. Our personal and political sovereignty are entirely subsumed under a responsibility to ensure we maintain a habitable planet. Those old hippies captured this idea perfectly in referring to the sovereignty of "Mother Earth." And this is where perhaps the most stinging criticism from the left has real bite: capitalism's near-total inability to value natural capital — economically, socially, or by any other method — either as an ongoing source of economic wealth (think, fish in the sea) or as the natural system that is the ultimate infrastructure supporting all human activity. The great investor and "inside" critic of current economic activity Jeremy Grantham[vi] has long lamented the absence of natural capital on our collective balance sheet. He writes, "Mainstream economics ignores natural capital. A true Hicksian[vii] profit

---

vi   Grantham is co-founder and chief strategist of GMO, a Boston-based global investment group that manages more than $100 billion on behalf of more than a thousand institutional investors. He's well-known in the investment community for predicting, and profiting from, many of the recent financial crashes and now provides leadership on climate risk through support of the London School of Economics' Grantham Research Institute on Climate Change and the Environment.

vii   John Hicks was noted for his formulation of the required balance between three commodities: money, consumption, and investment. A corporation is unwise, for example, to allow investors to consume dividends at the expense of investment that protects the long-term capital value of the company. Extending this idea to the full ecosystem that supports all economic activity is not much of a stretch, yet it continues to elude mainstream economic and corporate thinking. That Grantham is an outlier is, to my mind, a deep shame of working financial leaders.

requires that the capital base be left completely intact and only the excess is a true profit. Of course, we have not left our natural capital base intact or anything like it."[51] If the earth were a trust fund, we've long been spending the capital base.

The answer from some on the far right[viii] is full three-dimensional private ownership of natural capital (air, ground, and water!) on the assumption that private owners are motivated through self-interest to protect those natural systems to the extent they have value. Map our ecosystem, carve it up, and sell the pieces to the highest bidders — an ideological overreach of gargantuan proportions. Many on the far left vilify any market-based mechanism that seeks to mediate our interaction with natural systems as the same overreach in disguise. While I share concerns that the outright commodification of nature is deeply problematic, it's false to say all market mechanisms are some backhanded, sneaky privatization of the commons. Pricing externalities, like carbon emissions, for example, is a moderate's way to link market dynamics to natural value without the overreach of privatization.

The absence of natural capital in our accounts relates to a long-standing criticism of our standard measure of wealth generation. A country's GDP is not a good measure of long-term economic health. In each of the following real scenarios, GDP (as normally measured) nominally goes up: the Exxon Valdez spills a massive load of Alaskan oil along the American and Canadian northern coasts; fires rip through and around Fort McMurray in Alberta in the baking dry summer of 2016; industrial fishing fleets suck massive amounts of cod off the Newfoundland coast in an effort to maximize profits prior to the stocks' inevitable collapse. Natural disasters and unrestricted stripping of natural resources are seen as *good* things through the GDP lens, since both lead to short-term gain.

The pristine northern Pacific coasts have no immediate monetary value, they exist on nobody's balance sheet, not even the public's. Hence, their destruction goes undetected by normal economic yardsticks. Yet post-Valdez cleanup efforts required spending huge amounts

---

viii   Including Laura Jones of Canada's libertarian cheerleader, the Fraser Institute, and Fred Smith Jr., founder of America's equivalent, the Competitive Enterprise Institute.

of money, public and private, to deploy people and equipment, all of which adds nominal economic value. Same goes for the forests around Fort McMurray. The trees' destruction, unless already owned by a forestry company, cost nothing, but the planes flying to drop fire-retardant certainly did. The buildings destroyed had value, of course. But they were soon replaced, bringing a mini economic boon at the expense of insurance companies and the public purse.

And those disappearing cod had value only once brought onto a boat. In the water, living and breeding, they are effectively worthless. Yet economic devastation followed their disappearance, particularly for local Newfoundlanders who had neither the fleet nor inclination to overfish. Large Spanish and Russian factory fleets operating just off the 200-mile economic zone did most of the damage. Once done, they simply moved on. Incentives to cooperate and preserve stocks were overrun by naked self-interest and the limits of national waters. Then–Canadian minister of fisheries and oceans, Newfoundland-born Brian Tobin, lamented the factory boats capturing immature fish and made a show of confiscated nets at the UN. But to no avail.

The parallels from cod-stock collapse to climate risk are disturbingly clear: those with least capacity to damage the atmosphere will suffer the most harm. Those with the industrial might to lead change occupy themselves with existing paths to profit. Well-meaning politicians make earnest statements. The scientific evidence of risk is clear. The irrationality of what's happening is obvious. And so are the potential solutions. Yet on we go.

The common thread in these disasters is simple: without a price on something, our economic system is blind to its value, be it economic, social, or spiritual. The total stock of a remaining natural resource has no value unless it already sits on a corporate balance sheet. Only the flows — that which we take from the remaining natural stock — are measured and valued. Only what we extract has value, and only at the time of extraction. Unless we immediately place similar economic value on our ecosystem's continued stability, ongoing economic activity will soon burn its way to permanent instability at staggering long-term cost.

Capital expenditures to fund adaptation to climate disruption face a similar paradox. When we anticipate or react to a catastrophe, GDP goes up. Hurricanes Harvey and Sandy ripped through Houston and New York, devastating[ix] infrastructure and bringing ruin to countless lives. While there may have been a short drop in local economic output, the net result was an increase in national output as insurance companies, FEMA, and state and national coffers rebuilt those areas. It's hard to see how rebuilding existing capital stock, from streets to buildings, pipelines to refineries, adds to long-term economic security. We may be running our economic engines hard, but we're standing still.

Same goes for anticipatory action, like when we build resiliency — dikes, seawalls, more robust sewer water management systems, stronger bridges — to protect us from the coming storms. The cost of protecting New Orleans from further Category 5 hurricanes is estimated at $30 billion[52]; the massive pumping systems deployed in Miami Beach for half a billion dollars to (temporarily[x]) keep that piece of Miami from submerging under surging seas; moving telecoms and IT equipment to higher floors in New York in reaction to Hurricane Sandy; upgrading bridges to handle higher winds and water flows — all that work adds *nothing* to productivity but merely tries to protect what already exists. Like the Red Queen Alice encounters in *Through the Looking-Glass*, we are increasingly sprinting to stay in place. Those same funds can be spent in far more productive ways that add to long-term productivity.

The need for capital expenditures that anticipate climate disaster results in the first place from inadequate incentives to maintain a relatively calm status quo. We currently place little to no value on a functioning ecosystem. The crazy part is it costs way less to mitigate this risk than adapt to it. Remember: once we've kicked the climate into instability, we must react to that instability for the next several centuries! We're willing to spend money on big, expensive Band-Aids, but not to avoid the cuts and scrapes in the first place.

The libertarian approach to fixing this mess is to privatize everything: the North American coasts, the forests of Alberta, the fish in the

---

ix   Harvey caused an estimated $125 billion in damages, and Sandy caused $70 billion.
x   Miami can't be protected from rising seas by a sea wall, since it sits on top of porous, sponge-like ground, hence water seeps up from underneath.

ocean, and, yes, the atmosphere itself. It's easy to think of this as a kind of loony straw man, designed to provoke discussion, but not a serious proposal. Yet in Bolivia, privatization of water led to a situation where "licenses were even required for individuals to collect rainwater from their roofs, and people were charged for water taken from their own wells."[53] When some people speak about three-dimensional ownership, they are not kidding. There's good reason to beware libertarians bearing ecological gifts.

Yet, overreaction to this extreme, neo-Randian objectivist nonsense can lead to unwarranted rejection of pragmatic solutions that use market signals to link global objectives to individual action. Carbon pricing, and in particular cap-and-trade, has been vilified by some on the left as being just such libertarian hogwash in disguise. It's important to distinguish *privatization* of the commons (water, the atmosphere, land) from a *cost on interacting* with those commons. In his otherwise wonderful book, *Talking to My Daughter About the Economy*, Greek economist Yanis Varoufakis conflates the two, believing they are conceptually and inextricably linked:

> The same goes for the atmosphere. . . . If [this] were privately owned, then industries would be forced to pay for the right to emit pollutants into the air . . . ensuring [it is] used in moderation while the owner ensured they were protected and sustained. . . . The . . . atmosphere could just as easily be bought and sold in small pieces by thousands of different owners in markets designed especially for that purpose.[54]

Varoufakis's legitimate suspicion of private ownership of natural systems blinds him to the idea that commons, like the atmosphere, are not in danger of being privatized by market-based mechanisms that enforce compliance with publicly imposed limits on emissions. Cap-and-trade is no more stealth privatization of the skies overhead than speeding tickets are stealth privatization of our roads or an entry price to National Parks a privatization of those lands. Assigning a price to interact with the atmosphere is not the same thing as assigning ownership.

What Varoufakis and a number of other critics have right is that pricing carbon (or any other market-based value assignation to a natural system) implies a commodification of that natural system. When we put a price on a ton of emissions, we're saying there's a quantifiable value to the atmosphere's composition in the absence of that extra ton. In theory, it makes sense — it's called the social cost of carbon. But practically, it's nonsense. Those calculations are a mug's game. The models on which they're based cannot capture the complexity or precision to make the answer mean anything. Garbage in, garbage out. Further, I'd argue there's no way to place an appropriate value on a functioning ecosystem other than to say, "it's infinite." In which case the models break down.

I think dollars and cents is the wrong language. Qualitative statements are more appropriate for this kind of existential risk, but that criticism is more philosophical than practical. And since I've argued that we should leave our dogmas at the door in order to come up with workable solutions to the climate problem, that includes my philosophical allergy to quantitative models trying to link the economy to the environment. One doesn't have to swallow whole the quants' belief we can accurately commodify the atmosphere down to the granularity of a single ton of carbon dioxide to endorse pricing carbon. One merely needs to acknowledge that putting a price on carbon emissions incentivizes people to emit less — and that's enough for now. There's a hole in our accounts through which carbon leaks. Pricing carbon is one way to plug that hole.

I've argued throughout this book that carbon pricing is not enough at this late stage to sufficiently mitigate climate risk and that each country will have to decide independently what other policy initiatives are best suited to its own economic and political circumstances. But one thing a price on carbon can do is provide a single, transparent, and level playing field for all economies, enforced by existing institutions like the WTO.

Environmentalists have long positioned the Earth as the ultimate sovereign — our collective Mother, the nurturing context within which all economic (and political and social) activity is contained and supported. They are correct. We are not separate from the environment; we are a part of it. We are totally dependent on it and always will be, despite billionaire dreams of escape velocity. It's from whence we, and all past and future civilizations, come. A global price on carbon is the first economic recognition of that ultimate sovereignty to which we all, forever, owe fealty. It is, above all else, a sign of respect for Gaia.

PART 3:

WELCOME
to the
ANTHROPOCENE

"Few men realize that their life, the very essence of their character, their capabilities and their audacities, are only the expression of their belief in the safety of their surroundings."[35]
— JOSEPH CONRAD

Dear Rupert,

When I started writing this book, you didn't exist. Now you do, and you bring me unimagined joy. I love you in ways I never thought possible. You have opened wide the tap of my full humanity, in all its frailty and love. Obama once described having a kid as having "your heart outside your body." At the time, I had no idea what he was talking about, other than it sounded unwise. I do now, and it isn't. When your mom and I had you, I became fully human. You filled a hole in the center of my being I didn't even know was there. Thanks, Little Monkey!

That happiness is tinged by a dark cloud. By the time you're old enough to read this, it's likely we will have committed our planet to at least 2°C of warming, perhaps much more. I do hope it's not so much more — the 4°C or even 5°C that is the stuff of nightmares. Indeed, were I religious man, I'd pray we left your generation some optimistic glimmer of halting warming under 3°C. If we have not, then I'm afraid you and your peers will have to live through some fearsome times. I do not know how to equip you for what may be coming.

As you will know by now, I've been scared of climate disruption for some time. Before you came along that fear was tempered by a kind of intellectual and emotional distance. While I — like anyone who comes to terms with climate risk — went through a period of anger and grief, I was able to compartmentalize those feelings. What I saw happening all around me was experienced as a kind of anthropological catastrophe, not a personal one. Of course, I felt a deep empathy for the people affected — largely the world's poorest — as they got pummeled by extreme weather, pushed out of their homes by rising waters, and as they fled across borders in search of security. But I could retreat behind a kind of philosophical screen to blunt the despair I now so often feel when I think of the future. It's all so personal now.

Before you came along, I'd use mental tricks to keep the climate demons at bay. Most involved taking some ridiculously long view to visualize what we humans were up to in as broad a context as possible. I did that in order to diminish a sense of moral failure and loss. I understand some cultures are accustomed to thinking in very long time frames — the Chinese in 5000-year cycles, for example. On that view, perhaps we're just doing what every creature does: bumping up against the limits of our ecological niche, like foxes who grow in number until there aren't enough rabbits to eat. The population collapses, only to rise once again. But instead of a forest, our niche is the whole planet. Instead of teeth and claws, our tools are technology. Such an air of evolutionary inevitability masks a moral failure. We are destined to become lost in history, identified eons out as the Carbon People, whose lasting geological record can be found in the sediment under the civilization of some future peoples.

On an even larger view, human civilization itself is just one experiment among countless others. "Life will go on," I'd say to myself in the days and years before you came along. "Humans screwed up in the here and now, but life continues here and elsewhere!" I found solace in Stuart Kauffman's view, expressed best in At Home in the Universe, *that the complexity of life is an inevitable outcome in a universe like ours, governed by the kinds of physical laws we've identified. Energy and matter necessarily entail life: we are not chance caught on a wing, but a natural expression of the universe. We are the universe conscious of itself, and our emergence was inevitable. If it happened here and now, it's happening elsewhere and everywhere. How much can this human experiment matter in such a universal context?*

These are the consolations of a philosopher, one who finds a necessary distance from the messy world in a place of abstraction, reflection. But none of this is happening at arm's length anymore. It's deeply personal because it's your future. It's like before you came along, I was playing poker with fake money, plastic chips. Now the stakes are real. And it changes everything for me. But lots of people have kids. How could we possibly fail when we all might act on that deepest, most primal human force — to protect our children from danger?

As I look at you now, playfully banging a flute against the couch with a mischievous look in your eye, and try to visualize your world as you read this, the hard limits of my imaginative powers are laid bare. It's either something like the world I live in with some added hardships, or it's Mad Max. One a linear extrapolation of what I already know, corresponding to a childishly naïve view of climate risk. The other a movie's portrayal of an apocalyptic world, complete with outlandish costumes and nightmarish violence. Neither vision is helpful. Neither is a guide to what I might teach you, give you, show you to help you on your way there, whatever "there" is.

Perhaps that's one reason we have failed to keep the world safe: we can't picture in personal terms what higher levels of warming mean to us, to you and your generation. Or we safely categorize what visions we can conjure up as dystopian movies, and hence likely fiction. Our minds' imaginative limits filter out the most likely outcomes. Hence, we live in delusion. What we think about when we think about catastrophic warming are really the most optimistic scenarios — those which allow linear extrapolation from our world to yours. But this limited kind of collective worry is just the floor of climate risk. To unlock that most human of instincts to protect our children requires us to worry about the most likely climate risk, which is much worse. And that is precisely where our imagination fails us.

The temptation to retreat from the world is a strong one. With so much love in the house and so much chaos and fear emerging in the world, why not just pack it in? Put all those worries aside, move to the country with some goats and chickens. Play Nature Man with my little family in the Eastern Townships, let other people worry about this stuff. Just hunker down, prepare for the worst, put up walls, protect my own.

That time will surely come, Rupe, but it's not now. I promise you today, as I write these words, that I will keep my shoulder to the wheel. I'll do my best. I'll keep trying to make this world a little less hot in all the ways I can: personal, political, and professional.

I wonder about your relationship with nature in a degraded world. Humans aren't good at perceiving incremental change. Biologists call this shifting baselines. Each generation grows accustomed to a diminished eco-system and thinks it's normal. Your natural background is not mine. For me, it was a comfort. My mind always settled more easily on the rich, organic dynamics I saw in nature than against the universe of artifacts that is a city. The patterns I saw in a drop falling off a leaf amidst the smell of damp earth on Newfoundland's East Coast Trail, or the play of air on water reflecting a skyline of trees from the shores of Algonquin, these were interactive poems that spoke of us as part of something bigger, universal, mysterious. But benign, supportive. Meditating on these small patterns often resolved my internal conflicts. But nature is part artifact now, and certainly not benign. My poet's Gaia is gone, replaced by an angry one. That change happened so quickly in historical time — a generation, really. But so slowly we can't see it.

Over the past few weeks, a brave sixteen-year-old Swedish girl, Greta Thunberg, has grabbed the world's attention. Last year, she went on strike at her school — by herself at the beginning — because she didn't understand the point of getting an education that couldn't prepare her for the world she sees coming. She's started something; a few months back, it was tens of thou-sands of people, and last week over a million. Maybe this is where the real climate action is. Not economics, discount rates, or cleantech. But millions of young people taking to the streets and refusing to leave until we adults feel their panic. Perhaps people like Greta will be heroes to you as you grow up in a hot world. But I wonder: will they be heroes because they tried or because they succeeded?

An apology from one generation to another feels entirely inadequate. But maybe that's a good place to start: I'm so sorry we couldn't act decisively enough to keep the Climate Bear at bay. My only hope is we did enough that you've still time to avoid the worst. My fear is we did not. Please know many of us tried.

I love you, Rupe. It's over to you and your friends now.

Lots of love,
Your dad

THE ANTHROPOCENE: OUR BRAVE NEW WORLD

First popularized by climate scientist Paul Crutzen, the Dutch Nobel Prize–winning atmospheric chemist, the term Anthropocene refers to the newest geological epoch as one in which we humans play the dominant role. Following the Holocene, the 11,000-year warmish period since the last ice age in which humans evolved, the Anthropocene is a new geological epoch defined and shaped by us. Some argue it started before industrial civilization, when our ancestors transformed much of the Earth's surface to agriculture, or even before that, when our ascent to the top of the food chain invariably drove most of large fauna to extinction through hunting. Whether or not these lesser events define the start of the Anthropocene, it's clear that human carbon emissions are now the main factor that will set ecological conditions for us and our planet-mates for a very long time to come. What's this new epoch going to be like?

There are a couple of ways the Anthropocene might play out at this point. Staying within a 2°C "safe" boundary, where life goes on largely as before, is not one of the more likely outcomes. Regardless of what we do, climate risk will overwhelmingly determine the nature of our economic, social, and political systems from here on out. Without Climate Capitalism, the future economy will be defined by an endless war over resource scarcity, ongoing economic loss, and degraded infrastructure. With it, we can look forward to a radically rebuilt economy decoupled from carbon emissions and powered by a fully electrified clean energy system. In the former case, we all lose. In the latter, some lose and some gain. But either way, within a single generation, we will radically rewire the nervous system of the entire global economy. Either way, climate risk will be the single most important macroeconomic variable of life this century.

That we've actually come to this juncture — a choice between radical change enacted by us or imposed upon us — comes down to the frailties of human psychology as much as it does to the vagaries of any particular economic system (neoconservative free-market ideologues aside).

THE UNCHECKED ANTHROPOCENE: CLIMATE PORN

On the darker path — let's call it the Unchecked Anthropocene — carbon emissions remain on their current trajectory. The planet warms enough to unlock the more nightmarish positive feedback loops — like melting methane in the far north — which puts warming out of reach of any possible emission reduction strategy. The ice melting in places like Greenland becomes irreversible, ensuring that oceans will rise hundreds of feet. Food insecurity runs rampant. Migration levels overwhelm remaining goodwill, and migrants are met with violence at hard borders. The institutions and machinery that underpin modern civilization break down. Progress slows, then stops, and finally begins to reverse . . .

It's not hard to see how local catastrophes cascade into systemic global instability. Pakistan verges on starvation as glaciers feeding the world's largest irrigated agricultural area disappear. Their leaders hold developed cities hostage for food with nuclear tipped missiles.[i] Bangladesh's border with India is already militarized.[ii] As its low-lying land floods and its

---

i   See Gwynne Dyer's *Climate Wars* for a comprehensive account of the military and security implications of climate risk. No human group in history has starved without resorting to what military might they have. We should not expect modern states to act any differently, and the stakes are much higher for the global community.

ii   About a hundred people a year are killed by soldiers today as they try to jump razor fences.

citizens flee to high ground, that border becomes a killing field. As the American southwest reverts to desert, California dries out. Residents flee, adding to a financial crisis that started when Florida began disappearing underwater. Local militias fight over states' rights to water flows. Grain exports from Russia stop permanently (foreshadowed in the drought of 2012). U.S. exports plummet, as do Australian. No international food reserves remain. China's huge African land holdings keep them eating for a while, but local African populations begin a terror campaign to retake their food production.

All of this is possible in Rupe's lifetime. People alive today may see this future. This is not about our great-great-grandchildren. They'll have to come out the other side of this nightmare. Public discourse often refers to what might happen this century, but the warming we've set in motion goes on for millennia. It doesn't stop in 2100 just because our graphs do. All bets are off if atmospheric carbon levels rise toward 1200 ppm, as the clouds might disappear at that point, causing a massive rise of an *additional* 8°C.

An emerging genre of "climate porn" tries to paint the pictures hidden in the science. Clive Hamilton's *Requiem for a Species* made clear many years ago the mathematical inevitability of shooting past the safe zone of 2°C. *The Uninhabitable Earth*, by David Wallace-Wells, is truly terrifying. There have long been debates within environmental advocacy groups about the relative effectiveness of fear versus hope. Is it better to focus on optimistic outcomes we might achieve or on the bitter truth of the risks we face? My view is we need both. Fear can catalyze action, but only if it can be funneled into real, actionable positive outcomes. And those must be based on realistic hope. Otherwise, it's paralyzing.

The details of climate porn are not as important as its underlying theme. The risk of the Unchecked Anthropocene is existential: it threatens our existence as we know it. We face the destruction of much of what we take for granted as the fabric of modern civilization: reasonably robust civic institutions; the rule of law; a military and police force able to provide national and local security; and enough food, water, and energy for most of us to live what we'd hoped was a reasonably blame-free middle-class existence. Progress itself becomes an historic anomaly. So, too, will everyday carefree moments, like vacations where we worry about sand in our shorts and the size of our Christmas bonus. The Unchecked Anthropocene really is Mad Max. I find these ideas paralyzing as much as motivating.

Modern humans have faced existential threats before, of course. Indeed, when the nature of climate risk is pointed out in these kinds of superlative terms, eyes often roll, and people say, "Every generation faced risks they believed were existential and unique. And we always come through!" It's true we've faced down other existential threats: the nuclear confrontation of the Cold War is the most immediately comparable example. World Wars I and II would have seemed pretty existential for those caught up in what must have felt like unlimited unleashing of human violence. Perhaps our imaginations conjure another Black Plague, a global unchecked pandemic like Ebola.

But climate risk is different. A nuclear war can be averted by a few calm decisions. The two World Wars killed many, but the threat was never truly existential. When they ended, so did the horror. An unchecked pandemic is as much bad luck as ill-preparedness. But climate risk grinds on, slowly and surely. It creeps toward us — first glimpsed over the horizon, then spotted on a distant hill, then filling the horizon, to finally emerge plainly in sight, violent in demeanor. By then, it's too late. It's all-consuming, and it will not end. What was climate risk becomes permanent instability.

This is not a long shot. Our degree of certainty in the basic science is near-absolute. The effects of climate disruption are already accelerating past what were once worst-case scenarios. Still, we watch our emissions grow. All the while, we bicker and distract ourselves. The scariest thing to me was realizing no matter how imminent the threat or how clear its repercussions, recalcitrant, stupid, or sociopathic civic and political leaders still manage to poison public debate, rag the puck, delay, delay, delay. Houston can be flooded, California on fire, and the jet stream disrupted, yet climate policy remains a controversial wedge issue. Trump is not the only one to show how effectively "fake news" can drive public perception, although he is uniquely bald-faced. I always thought once things got bad enough, we'd inevitably act. I no longer believe that.

Catastrophic climate risk is unlike other existential threats, because its probability is not low, but high. A worst-case outcome is not predicated on bad luck or lousy negotiation, but on the mere passage of time. The Unchecked Anthropocene is our *default* future; it's what we get by doing nothing. It's our shared destiny if we just remain distracted for another decade or two. We readapt to ever-worsening baselines until at some unidentifiable point — like when a baby becomes a little boy or a little boy a young man — the problem has grown from background noise to a nasty fight. A fight over sharing ever-shrinking resources between ever-more-desperate people essentially forever.

It's not impossible that in an Unchecked Anthropocene, the earth shakes human civilization off its back. I doubt the human species will disappear entirely; we're far too resilient for that. But it's not as crazy as its sounds to say we are bringing about the endgame of a recognizably modern human civilization. In the distant future, we may be remembered as the Carbon People, named after our mark in the geological record.

We moderns are utterly unprepared psychologically for the role we must take on to avoid this fate. We need a new theology by which we have the responsibility — over *a single lifetime* — to keep alive the possibility of human progress. That is a far more mind-bending reality to confront than the mere death of a deity, as Friedrich Nietzsche pronounced. Or the moral emptiness of the existentialist universe that emerged after the evils of World War II, as proclaimed by Simone de Beauvoir and Jean-Paul Sartre. This is not just another "meaning of life" problem, but an "existence of life" problem.

The world order in an Unchecked Anthropocene is a horrid affair, looking something like Haiti in its darker moments of history. The wealthy hide fearfully in enclaves, much as in Haiti, where they lived behind armed guards on a hilltop, believing they can protect themselves from those below. If this sounds extreme, it's playing out right now. Uber-rich "preppers" are already setting up remote compounds to guard against the worst. New Zealand became a favored escape hatch for so many (including neocon zealot Peter Thiel) that it was forced to push back with tax and citizenship reform. But one cannot escape to an island.

We're too interconnected for that. And the crowds around the hill can't be expected to play nice.

Long before these worst-case scenarios play out, it will be clear enough to the global citizenry what's happening. Trump's brand of populism arose from the distrust, fear, and disenfranchisement that came from a mild recession in an era of gentle globalization. Whatever populism emerges from a global citizenry that faces food shortages, permanently eroded landscapes, and mass migration will not be benign. The misguided hope that we'll all come together will play out against a scramble for ever-more-scarce resources. To say nothing of the collective anguish of knowing there's no end in sight.

People will look for scapegoats. Right now, business leaders, Wall Street executives, and capitalism itself look like pretty good targets. Naomi Klein's venom against economic elites is infectious. People will not countenance nuanced arguments about growth rates, the difference between libertarian capitalism and social democratic systems, the necessity of global supply chains to minimize cost, and so on. They will be angry and scared. Those feelings will express themselves in ways we cannot predict. The populist backlash will not go well for those captains of industry most able to mitigate this risk.

If I have one message to elites, it's this: get on the right side of history on this issue, and do it now. Acknowledge the problem. Make a good-faith effort to be a part of the solution. Even if that means leaving money on the table in the form of high-carbon assets stranded by progressive policy. Indeed, fight for that policy! Sacrifice some wealth. If that means endorsing what is today called "socialism," so what? Capitalism must be radically reformulated — now — in ways that are clear, open, and visible to the public in order to defuse the certain existential risk of ongoing climate disruption. Political populism and public sentiment will not go your way in the Unchecked Anthropocene. And anyway, how much fun can it be hiding out in some high-tech compound stocked with five years of canned food and an airstrip?

Progress Curtailed

The Unchecked Anthropocene negates our deepest cultural norm: a sense of progress. Progress lies at the heart of modern life. Sir Francis Bacon declared back in the seventeenth century that with opposable thumbs and a rational mind, nature is ours to control. Since then, our personal, economic, political, and cultural lives have been underwritten by a sense of progress. Our very self-conception is based on the belief that the future is better than the past, that our children's lives will be an improvement over our own. It's why many of us get out of bed in the morning. Most of economics falls apart without progress (represented as both technological change and continued growth). Politics is all about promising a better future. Western culture is predicated upon the promise of progress.

Life without it is literally unthinkable. Absent a sense of progression over time, we barely know how to think about ourselves. The economist John Gray put it this way: "Even those who nominally follow more traditional creeds rely on a belief in the future for their mental composure. History may be a succession of absurdities, tragedies, and crimes, but — everyone insists — the future can still be better than anything in the past. To give up this hope would induce a state of despair."[56] For Gray, progress is an illusion. Scientific knowledge and technological prowess may increase, but they rest in the hands of we humans who retain the frailties of cognitive structures inherited over eons of evolution. Ethical and political progress are entirely dependent on institutions — political, social, legal — which are human affairs and prone to collapse. Without them, increased knowledge and advanced technologies are of little use. For Gray, however self-aware and in command of our destiny we believe ourselves to be, we're acting out an ancient deterministic evolutionary game. Like all animals, we bump against ecological limits. And when we do, it all comes crashing down. I hope an Unchecked Anthropocene doesn't prove Gray right.

That's why climate porn doesn't make for the best motivational speeches. It's hard to take in — literally. Because it doesn't square with how we see ourselves — our mental antibodies, as it were, reject arguments predicated on climate porn. Our unconscious engine is incompatible[57] with these kinds of beliefs. Since birth, our brains are busy wiring one idea to another, forming a vast and automatic set of connections that together form our worldview. Those connections form our uniquely human common sense, deployed constantly to make sense of the world around us. Our minds need this hardwired, unconscious worldview to function. New beliefs run the gauntlet of that worldview. Those counter to it are rejected.

Climate porn is an affront to that worldview. At the heart of our place in the world is the idea of progress: the future is better than the past; the economy can grow forever; human ingenuity knows no bounds; technology can solve our problems; nature is ours to control. We hold these beliefs — explicitly if asked and implicitly if not — because they reflect our experience. They permeate our culture from the fairy tales we read as children to the courses we study in university. And they have served us well. Until now. An Unchecked Anthropocene runs counter to these shared, deeply cherished ideas. We cannot hold in our minds both our existing worldview and the possibility of catastrophic climate disruption. It brings cognitive dissonance, uncomfortable feelings of anxiety. There's a reason people get so angry about climate.

Our minds have been playing tricks on us for decades to keep these disastrous possibilities at bay. You can measure this stuff.[58] We start by rolling our eyes in disbelief because our unconscious feels discomfort. We seek evidence to confirm that disbelief (confirmation bias), which permeates the internet and commercial TV. We refuse to accept evidence and arguments to the contrary, because we so dislike the conclusion (affect heuristic). We favor the views of those who think like us (peer group bias). And we allow nice, reasonable people to convince us of a nicer, more reasonable view of climate risk (halo effect). In trying to preserve an unconscious and optimistic view of ourselves in the world, we make possible the destruction of that very worldview.

There's good reason we've collectively failed to face up to these increasingly obvious dangers. It's human to do so. That doesn't make those dangers any less real. But it does mean we must be strategic about communicating these risks. Climate porn is necessary (as all truths are necessary), but it's not sufficient. We also need a way out of this mess, a compass to guide our mind as we grapple with climate risk in a realistic way. We need a story that speaks to the best in us, that corroborates the best of how we see ourselves. Only upon that kind of story might we motivate ourselves to build a different future. And that's what Climate Capitalism is all about: radically reforming our economic system to ensure that we can continue to evolve and improve our position in this world. Progress *does* make for good motivational speeches.

CLIMATE CAPITALISM: THE MANAGED ANTHROPOCENE

This book is not about hope, but about trying. Giving it a shot. Hope is nothing without effort. And if we really heave on the oars — if business leaders abandon the safety of incremental change, if citizens demand action in ways that cannot be ignored, and if politicians articulate what true leadership looks like — we might just manage this risk. We must abandon the old fight between left and right in favor of a pragmatic, centrist way forward. We urgently need to exchange entrenched positions for ready solutions. There's no time left to argue about what the best option is, to finesse our response. But there is just enough time to *try them all.*

When the United States got dragged into World War II, they didn't just stop avoiding the issue — they went all-in. From zero to maximum speed with one Japanese attack. No more debates about debt levels, about whether or not governments should drive industrial policy, or whether this new industrial posture was called "capitalism," "socialism," or any other-ism; the U.S. went to war and damned all who got in the way. That's the attitude we need now, in every country. Those who don't want to come along must be persuaded or forced. Climate risk calls for nothing less than an all-out global industrial push. That, we can do. Let's call it the Managed Anthropocene, in which we limit warming to well under 3°C. While less ambitious than the Paris accord's "official" goal of 2°C, it's still no picnic to achieve. But perhaps it's an Anthropocene we can live with. What does that kind of Climate Capitalism look like, and how might it come about?

It all starts a long way from Wall Street, Bay Street, the City, Capitol Hill, or Parliament Hill. A little girl in Sweden, Greta Thunberg, stages a solo climate strike outside the Swedish Parliament in 2018. She's frustrated because she doesn't understand why she should spend time on an education that won't matter in the world she sees coming. Support for Greta and her profile as an emerging climate leader grows in leaps and bounds. Initially opposed by sober-sounding vice principals and even a few prime ministers, the student strike goes global throughout 2019 (at the time of writing, the #FridaysForFuture movement had passed the million mark for the first time, and feels unstoppable).

And we can envision where we might go from here in a world that adopts Climate Capitalism. By 2020, many millions of students leave school one Friday each month, expressing their fear and outrage in ways adults cannot ignore. Instead of us teaching them, they teach us. Greta is named UN Special Envoy on Climate and wins the Nobel Peace Prize. General strikes begin in the developed world throughout 2021. By the end of that year, millions of regular adult citizens join the students to sit down peacefully one Friday a month. Cities shut down. Those who ridicule or harass the strikers are vastly outnumbered by those who support them. People get comfortable publicly expressing fears that, until then, were private. The strikes spread to the developing world, with the exception of China, who brutally cracks down on "climate dissidents." Global climate consciousness is fully awake now. And the political will we long needed has finally lit up, sparked by a girl too young to drive.

Going further down this road, business leaders who won't commit themselves and their companies to deep emissions targets find they can't retain talented employees. Even investment banks on Wall Street have trouble attracting their usual cadre of Ivy League graduates with promises of big money. Shareholders demand, and get, comprehensive executive compensation reform: salaries and bonuses are linked to emission targets, which no longer seem merely aspirational. Incumbent industries, like heavy-oil producers in Canada, learn the hard way that their social license to expand operations is gone.

Carbon risk disclosure becomes the norm by 2020. Shareholders begin to exit fossil fuels stocks a few years later, in a panic as the carbon bubble pops with a bang. Capital gets scarce, hence expensive, for any long-term project that's anything but zero carbon. Climate risk disclosure ramps up in the mid-2020s as bottom-up models get better at articulating climate risk in financial terms, with actionable advice. Pension funds respond to their membership and commit as a group to prioritize capital for projects that do more than reduce their own carbon risk but go further than that to mitigate climate risk itself.

It's too late for some investors. Swaths of Miami and California real estate converts to a collective Real Estate Investment Trust at massive discounts to previous value. Short-term rentals become the norm, and the first large-scale middle- and upper-class climate migrants push up real estate prices in places like Chicago, Seattle, and even Toronto. Canada becomes a hipster destination of choice for climate converts. Court cases are fought over compensation for coal and natural gas plants that were forced to shut by a renewed regulatory zeal coupled with competitive clean energy systems. Canada's nationalized Trans Mountain Pipeline project becomes a globally ridiculed white elephant. But lots of winners emerge: heat pump manufacturers become a surprise entrant as one of the most valuable industrial stocks, as do energy storage developers and suppliers.

Politicians follow the lead of their citizenry. Renewed global leadership from a revitalized and assertive United States Congress forces the hand of the World Trade Organization. Countries that do not aggressively price carbon and meet ever more stringent emissions targets face steep tariffs on their exports. The World Bank allocates more than $5 trillion to low-carbon infrastructure, although a lot of capital starts to bleed off into resilience projects like seawalls, flood protection, and macro-scale irrigation systems. International agreements are formed to ensure adequate food reserves are allocated by a new and mighty UN Food Security Agency.

Clean energy infrastructure rolls out in three distinct waves. From 2020 through the middle of the decade, it's more of the same. But lots more. The solar superhighway built on Chinese manufacturing might spreads to other countries, because the processes are now largely automated and require more capital than people. Offshore wind turbines get bigger. Pension funds dabble in new project development instead of just buying up already operating assets. That fires up financing available to build new projects, and clean energy developers become the largest-growing global employers.

Small- to mid-scale distributed energy systems spread like wildfire, fueled by a slashed regulatory burden. Huge efficiency gains are realized across most industrial sectors, driven by strong AI, sensors, and automation. Energy storage is still mainly small(ish) battery systems, which can't keep up with the growing proportion of intermittent renewables like solar and wind. Conservatives howl about potential grid collapse. The popularity of electric cars far outstrips rates of production. Waiting lists are more than two years for any all-electric model by the mid-2020s. China looks like it intends to dominate electric vehicle production just as it did traditional solar. Alberta's heavy-oil production stays flat, as does global oil demand. The global cleantech market, however, grows at a healthy clip to more than $2 trillion annually by 2025. Canada's share of that market makes cleantech the single-largest aggregated industry in the country. Global carbon emissions peak in 2022.

Starting in the mid-2020s, another wave of more advanced tech rolls out at huge scale: new kinds of grid-scale energy storage, like compressed air and flow batteries that can operate for days; distributed solar and new storage now outcompetes any other fuel source as reliable baseload power; high-performance optics that combine light management, super-high-efficiency solar cells, and controllable thermal loads are integrated directly into building envelopes to form a new kind of urban utility market; the first profitable large-scale cellulosic fuel plants are commissioned, diverting forestry and agricultural waste into aviation fuels; the abandoned DESERTEC project that links north African solar farms to European grids is resurrected.

Utilities and fuel producers forced to abandon fossil fuel assets before the end of their operating life are no longer given direct compensation. Instead, they're offered subsidized risk capital to deploy these next-generation technologies at scale, allowing them to leverage their remaining (and aging) engineering bench strength and market access. Those that refuse to adapt shut down. Everything is a target for electrification — cars, buses, industrial and residential heat. The more nimble utilities scramble to find business strategies that allow them to offer energy-as-a-service, where it's not energy itself that matters, but what it can accomplish: the temperature of a building, for example, not how it's heated or cooled; or getting from A to B as quickly and comfortably as possible, instead of caring about how many horses are under the hood.

Trade rules are relaxed to allow countries to subsidize these deployments of advanced cleantech at scale, which ignites huge competition for market share across the globe. At first, innovation came from the usual places in developed countries, but soon India and China are able to bring their engineering talent to the table. As a result, manufacturing is better distributed, and China begins to respect intellectual property rights.

Electric cars are now self-driving — essentially computers on wheels. Individual ownership plummets as cars become shared assets, allowing fully electrified auto manufacturers to finally meet growing demand. Those who relied too long on the internal combustion engine, however, go bankrupt almost overnight. Places that subsidized their factories see employment plummet. Cities become more livable as all that space for parked cars is recaptured. Congestion eases. Oil demand has long peaked and is headed down. The price of oil collapses, anticipating a long and irreversible ride down the price slope. Alberta's heavy-oil market disappears, and the province heads into a deep recession. The cleantech market grows to $7 trillion a year. Global carbon emissions are dropping by more than 3 percent annually.

Then, from 2040 onward, even deeper disruptions are finally brought to market to capture the most recalcitrant remaining emissions. Modular molten-salt reactors support concentrated centers of industrial heat and power, using reformed spent fuel rods. Nuclear waste becomes nuclear fuel. Enhanced geothermal wells are dug next to any remaining natural gas plants to replace baseload power requirements. Short-haul flights are now electric, longer-haul jets run mostly on cellulosic fuel. Unemployment due to automation is offset by an ongoing multi-decade effort to retrofit building stock with energy efficiency measures.

Markets for carbon pricing have matured significantly. Cleantech was competitive enough that the price never needed to go above $100 per ton to displace fossil fuels. But that pricing mechanism was refined to allow countries like Ecuador, Brazil, and Indonesia to get financial compensation in return for protecting their forests. Deforestation grinds to a halt. Conservative howling about "payments for nothing" is met by social media mobs distributing photos that parody billionaire offspring, the highlight being an obese Donald Trump, Jr. asleep on a solid gold toilet. Areas of marginal land are planted with switchgrass, which quickly becomes the most viable way to sequester carbon. The cleantech market exceeds ten trillion. By 2050, we have not yet reached net zero emissions, but the target is clearly within sight.

None of this is rocket science. Most of what's needed has already been invented. What remains is commercialization and deployment. None of this is particularly difficult or risky. All it takes is will and effort. Perhaps that little Swedish girl really is the start of something big. For the sake of my little Rupe, I sure hope so.

ENDNOTES

1   Arnold J. Toynbee, *A Study of History: Volume I, Abridgement of* (London: Oxford University Press, 1987), 570.

2   PriceWaterhouseCoopers annual survey 2016, as quoted in "Davos 2016: Worries Mount for World's Business Leaders," *The Guardian*, January 19, 2016.

3   Gwynne Dyer, *Growing Pains: The Future of Democracy (and Work)* (London: Scribe Publications, 2018), 195.

4   All quotes in UNFCC press release: "Historic Paris Agreement on Climate Change: 195 Nations Set Path to Keep Temperature Rise Well Below 2 Degrees Celsius," UFNCC News, December 13, 2015. With the exception of Ban Ki-moon, which was from Twitter.

5   Climate Central, "Earth Flirts with a 1.5-Degree Celsius Global Warming Threshold," *Scientific American*, April 20, 2016.

6   Adrian E. Raftery, Alec Zimmer, Dargan M.W. Frierson, Richard Startz, and Peiran Liu, "Less than 2 °C warming by 2100 unlikely," *Nature Climate Change* 7 (July 2017): 637–641, doi.org/10.1038/nclimate3352.

7   Glenn A. Jones, Kevin J. Warner, "The 21st Century Population-Energy-Climate Nexus," *Energy Policy* 93 (June 2016): 206–212, doi.org/10.1016/j.enpol.2016.02.044.

8   Joe Romm, "Misleading U.N. Report Confuses Media on Paris Climate Talks," ThinkProgress, November 3, 2015, thinkprogress.org/misleading-u-n-report-confuses-media-on-paris-climate-talks-7cc9ef239328/. Also see the Climate Scoreboard by Climate Central (climateinteractive.org/programs/score-board/), which is calibrated to the IPCC Fifth Assessment Report.

9   Naomi Klein, "The Carbon Tax on the Ballot in Washington State Is Not the Right Way to Deal with Global Warming," *The Nation*, November 4, 2016.

10   Ibid.

11   Response in a tweet.

12   Ibid.

13   Naomi Klein, *This Changes Everything: Capitalism vs. The Climate* (Toronto: Knopf Canada, 2014), 54–55.

14   Ibid., pg. 22

15   Ibid., pg. 21

16   Gwynne Dyer. *Growing Pains: The Future of Democracy (and Work)* (London: Scribe Publications, 2018), 195.

17   Ibid., pg. 23

18   Ibid., pg. 23

19   Ibid., pg. 21

20   Ibid., pg. 21

21   Ibid., pg. 18

22   Ibid., pg. 25

23   Regarding marginal tax rates, see Eric Levitz, "Poll: Majority Backs AOC's 70 Percent Top Marginal Tax Rate," Intelligencer, *New York Magazine*, January 15, 2019. Regarding gun laws, see Steven Shepard, "Gun control support surges in polls," *Politico*, last updated February 28, 2018. Regarding health care, see Yoni Blumberg, "70% of Americans Now Support Medicare-for-All—Here's How Single-Payer Could Affect You," *CBNC Money*, August 28, 2018.

24   Michael Shellenberger and Ted Nordhaus, *Break Through: Why We Can't Leave Saving the Planet to Environmentalists* (San Diego: Mariner Books, 2009), 15.

25   Ibid., pg. 17

26   Ibid., pg. 38

27   Ibid., pg. 120

28 All quotes in this section from "An Ecomodernist Manifesto" (multiple authors) unless noted otherwise. Available online at ecomodernism.org.

29 E.O. Wilson, *Half-Earth: Our Planet's Fight For Life* (W. W. Norton & Co, 2016), 241.

30 Pieter Gagnon, Robert Margolis, Jennifer Melius, Caleb Phillips, and Ryan Elmore, "Rooftop Solar Photovoltaic Technical Potential in the United States: A Detailed Assessment," The National Renewable Energy Laboratory (NREL), January 2016, nrel.gov/docs/fy16osti/65298.pdf.

31 Michael Shellenberger and Ted Nordhaus, *Break Through: From the Death of Environmentalism to the Politics of Possibility*, (San Diego: Mariner Books, 2009): 225.

32 Francis Fukuyama, *The End of History and the Last Man* (New York: The Free Press, 1992), afterword.

33 Jacobson's original article appears to have been withdrawn from the Stanford University site. See instead, "Critique of 'A Path to Sustainable Energy by 2030,'" Brave New Climate, bravenewclimate.com/2009/11/03/wws-2030-critique/.

34 Unless otherwise noted, all quotes from Carney's speech to Lloyd's of London: Mark Carney, "Breaking the Tragedy of the Horizon — Climate Change and Financial Stability." Speech, London, September 29, 2015. bis.org/review/r151009a.pdf.

35 Source: Bloomberg New Energy Finance.

36 "The Outlook for Energy: A View to 2040," ExxonMobil, 2016, cdn.exxonmobil.com/~/media/global/files/outlook-for-energy/2016/2016-outlook-for-energy.pdf.

37 As quoted in, Arthur Nelsen, "Climate Change Could Make Insurance Too Expensive for Most People — Report," *The Guardian*, March 21, 2019, theguardian.com/environment/2019/mar/21/climate-change-could-make-insurance-too-expensive-for-ordinary-people-report.

38 Salvatore Basile, author of *Cool: How Air Conditioning Changed Everything*, quoted in, Rory Carroll, "How America Became Addicted to Air Conditioning," *The Guardian*, October 26, 2015, theguardian.com/environment/2015/oct/26/how-america-became-addicted-to-air-conditioning.

39 "2011 World Energy Outlook," International Energy Agency, 2011, iea.org/publications/freepublications/publication/WEO2011_WEB.pdf.

40 Ibid.

41 Jennifer A. Dlouhy, "Yellen Touts Carbon Tax as 'Textbook Solution' to Climate Change," *Bloomberg News*, September 10, 2019, bnnbloomberg.ca/yellen-touts-carbon-tax-as-textbook-solution-to-climate-change-1.1135433.

42 Joseph A. Schumpeter, *Capitalism, Socialism and Democracy* (Routledge, 1994), 82–83.

43 Tom Harries and Meredith Annex, "Orsted's profitable transformation from oil, gas and coal to renewables," Bloomberg New Energy Finance, December 2018, poweringpastcoal.org/insights/economy/orsteds-profitable-transformation-from-oil-gas-and-coal-to-renewables.

44 See corporate website, equinor.com.

45 Adam Vaughan, "Shell would support UK bringing forward petrol ban from 2040," *The Guardian*, July 5, 2018, theguardian.com/business/2018/jul/05/shell-would-support-uk-bringing-forward-petrol-ban-from-2040.

46 Chris Hayes, "The New Abolitionism," *The Nation*, May 12, 2014, thenation.com/article/new-abolitionism/.

47 Toby Heaps, "What is your Climate Change Business Plan?," *Financial Times*, May 31, 2015, ft.com/content/f5c6b73e-0390-11e5-a70f-00144feabdc0.

48 Tim Flannery, *Atmosphere of Hope: Solutions to the Climate Crisis* (New York: Atlantic Monthly Press, 2015).

49 See the full report here: "Report from Canada's Economic Strategy Tables: Clean Technology," The Government of Canada, last modified October 4, 2018, ic.gc.ca/eic/site/098.nsf/eng/00023.html.

50 Matthias Kroll, "We print money to bail out banks. Why can't we do it to solve climate change?" *The Guardian*, January 30, 2016, theguardian.com/global-development-professionals-network/2016/jan/30/print-money-climate-change-green-bond-quantitative-easing.

51 Jeremy Grantham, "The Race of our Lives Revisited," GMO White Papers, August 8, 2018, gmo.com/americas/research-library/the-race-of-our-lives-reinvested/.

52 Stéphane Hallegatte and Patrice Dumas. "Adaptation to Climate Change: Soft vs. Hard Adaptation," OECD, 2008.

53 Lisa Boscov-Ellen, "Thirst for Profit: Corporate Control of Water in Latin America," Common Dreams, Council on Hemispheric Affairs, June 2009, commondreams.org/views/2009/06/20/thirst-profit-corporate-control-water-latin-america.

54 Yanis Varoufakis, *Talking to My Daughter About the Economy: or, How Capitalism Works—and How It Fails* (New York: Farrar, Straus and Giroux, 2018).

55 Joseph Conrad, "An Outpost of Progress," *Cosmopolis* (July 1897).

56 John Gray, *The Silence of Animals: On Progress and Other Modern Myths*, (New York: Farrar, Straus and Giroux, 2013), 4.

57 Tom Rand, "Chapter 2: The Siren Song of Denial," in *Waking the Frog: Solutions for Our Climate Change Paralysis* (Toronto: ECW Press, 2015).

58 Ibid.

INDEX

Footnotes are denoted by **n** *following the page number.*

Powerwall, 65
Beta Renewables, 66
bio-energy with carbon capture and
storage (BECCS), 10 nIV, 117–118,
153
Black Hats. *see* corporate leaders
Bombardier, 30, 178
*Break Through: From the Death of
Environmentalism to the Politics of
Possibility*, 26–29
Breakthrough Energy Coalition, 3
Breakthrough Institute, 26–29, 40
breeder reactors, 149 nxxvi, 151
Brexit, 186
British Columbia, carbon tax in, 12,
18, 89
Business Development Bank of
Canada (BDC), 162, 173
buy local provisions in trade
agreements, 187, 190

Cambridge Analytica, 113 niii
Canada
diplomatic strengths of, 5, 7
energy incumbents in, 82
focus on exports, 163–164
low-carbon sector in, 67, 160–161
need for multi-jurisdictional
climate governance in, xxxiii
need to show leadership on
climate change, xxx, 5
Canada Green Bond, 174–176
Canada's Oilsands Innovation
Alliance (COSIA), 150
cap-and-trade, 88–89, 108–109, 196
capital expenditures (CAPEX), vs
operating expenditures (OPEX),
143–148, 162, 172
capital markets
bonds and, 176–177

as conservative, 173–174
co-opting of, xxix
capitalism
acknowledging flaws of, 54–55
characteristics of, 51–59
economic growth and, 26, 50–51,
59–75
government interference and, 52
as mixed blessing, xxvii–xxviii
moral status of, 51
need for reform of, 210
overview of, xxi–xxii, 49–51
public sector and, xxxii
unregulated capitalism, xxviii,
xxxiv nx
variants of, 94–95
carbon, social cost of, 197
carbon bubbles, xxvii, nix, 44, 60,
123–129
carbon capture and storage (CCS)
criticisms of, 152
as essential, 155–156
exclusion of, 121 nx
history of, 149
methods for, 152–153
public funding for, 150
scaling of, 40–41
carbon dioxide (CO_2)
capture of
. *see* bio-energy with carbon capture
and storage (BECCS); carbon
capture and storage (CCS)
myth of benefits of, 154 nxxxii
storage of
. *see* carbon capture and storage
(CCS)
carbon emissions
Anthropocene and, 205–206,
216–217
capture of, 154

trade agreements and, 188
climate agreement, 188–189
Climate Awareness Bond (Europe), 175
Climate Bears, 59, 68–70, 205
Climate Capitalism
    carbon pricing and, 79–83, 88–111
    as centrist, 47
    and development of alternative
        energy sources, 69
    green growth and, 50–51
    mapping of, 46–49
    need for speed and, 84–88
    as new set of economic rules, xxii
    as pragmatic, xvi, xix–xx, 46
    state as critical partner in, 69
    use of existing capitalist
        framework, 20–21, 48
climate change
    description of the problem, 15–46
    Paris Agreement, 7–15
    as population issue, xii
    social upheaval and, xiii
    as threat multiplier, xiii
    . *see also* climate disruption
climate deniers, xiv, 107
climate disruption
    conflicting positions on, xix–xx, 6
    cultural divide and, xxxiv
    emotional distance from, 202–205
    as existential threat, 208
    as having no borders, 181–182
    link to economic growth, 26
    Margaret Thatcher on, xxvi–xxvii
    Mark Carney on, 43–46
    relationship to global economy,
        15–16
    . as a "tragedy of the commons" or
        "tragedy of the horizon," 43
    use of term, xxii–xxiii
    . *see also* climate change

Climate Interactive, 11
climate options, ranking method for, 87
climate policy
    carbon pricing and, 80
    as wedge issue, 18–19, 87, 208
climate porn, 206–212
climate risk
    assessment of, 130–131
    as determining systems of the
        future, 205–206
    as existential, 57
    financial risk, 44
    liability risk, 44
    Mark Carney on, 44, 47–48
    peer-reviewed evidence and, 37
    as permanent instability, 208
    as philosophical issue, xii–xiii
    physical risk, 44
    as threat multiplier, xxiii
    as a "tragedy of the commons" or
        "tragedy of the horizon," 45
    underestimation of, 36
    . *see also* carbon risk
climate science, 130
Climate Service, 131–132
climate trauma, xi ni
Climeworks, 154
clouds, disappearance of, 156, 207
Club of Rome, 59, 70–75
CO_2. see carbon dioxide (CO_2)
coal
    apologists' view of, 38–39
    carbon content of, 125 nxiv, 128
    "clean" coal, 40
    competing with, xxi
    transition from, 33, 116–117, 141
cod, disappearance of, 74, 194
cognitive dissonance, 212
command-and-control policy, xxii ni,
    140

emissions
  agriculture and, 81
  forest management and, 80–81
  as pervasive to economic activity, 12
  social behavior and, 81
emissions targets
  1.5°C target, 8–10, 110
  2°C target, 10–13, 201, 205
  3.5°C target, 13–14
  mandatory, 69
  voluntary targets, 81
energy efficiency
  innovations in, 64
  paybacks from, 83, 147–148
energy giants
  absence of in cleantech space, 33
  as forced actors, 115, 122
  innovation and, 118–119
  lack of drive for innovation, 122
  as leading actors, 118–119
  rethinking role of, 64–65
  as willing actors, 116
energy plants, turning off of, 127
energy storage
  compressed air and, 166–167
  distributed energy storage, 148–149
  future of, 216
  solving problem of, 65
energy storage markets, 158
energy systems
  payback periods for, 145
  as slow to change, 33
  using capital budgets for, 146–147
energy-saving retrofits, 89, 144–147
enhanced geothermal systems (EGS), 149–151, 217
Epstein, Ken, 36–38
equilibrium, 90–95
Equinor (Statoil), 119–122, 128–129

Equinox Blueprint: Energy 2030, 150–151
Equinox Summit (WGSI), 150–151
ethanol, 66, 141–142
Evok Innovations, 122
evolutionary economics, 88, 90, 96, 98
Export Development Canada (EDC), 162–164, 173
externalities, 79, 89, 92, 193
Exxon Valdez, 193–194
ExxonMobil, 33, 112

Fabius, Laurent, 7
Facebook, 113–114
fear vs hope, 207
feed-in tariffs (FITs), 62, 159, 170
Fenix Energy, 145 nxxiv
fiduciary obligation, 45, 133–137
Figueres, Christiana, 7–8, 69
Flannery, Tim, 152
flexible regulations (flex-regs), 87, 101, 139–142
  . see also regulations
flooding, cost and probability of, 131
Ford, Doug, 89, 108–109, 142
forest management, 80–81
Fort McMurray, Alberta, 193–194
fossil fuel apologists, 4, 35–40
fossil fuel fatalists, 32–35
fossil fuel reserves
  digging up all of, 123
  net present value (NPV) of, 125
  relative carbon content of, 128–129
  stopping spending on, 129
  stranding of, xxvii nix, 127, 183, 210
  valuation of, 125–126
"fossil of the year" award, 7
Fraser Institute, xxxi, 4, 36, 38–39, 107
free trade, 52, 186–187, 189
freedom, concept of, 52